Developments in Primate Gesture Research

Gesture Studies (GS)

Gesture Studies aims to publish book-length publications on all aspects of gesture. These include, for instance, the relationship between gesture and speech; the role gesture may play in social interaction; gesture and cognition; the development of gesture in children; the processes by which spontaneously created gestures may become transformed into codified forms; the relationship between gesture and sign; biological studies of gesture, including the place of gesture in language evolution; and gesture in human-machine interaction. Volumes in this peer-reviewed series may be collected volumes, monographs, or reference books, in the English language.

For an overview of all books published in this series, please see *http://benjamins.com/catalog/gs*

Volume 6

Developments in Primate Gesture Research
Edited by Simone Pika and Katja Liebal

Developments in Primate Gesture Research

Edited by

Simone Pika
Max Planck Institute for Ornithology

Katja Liebal
Freie Universität Berlin, University of Portsmouth
& Max Planck Institute for Evolutionary Anthropology

John Benjamins Publishing Company

Amsterdam / Philadelphia

∞™ The paper used in this publication meets the minimum requirements of the American National Standard for Information Sciences – Permanence of Paper for Printed Library Materials, ANSI Z39.48-1984.

Library of Congress Cataloging-in-Publication Data

Developments in primate gesture research / edited by Simone Pika, Katja Liebal.

p. cm. (Gesture Studies, ISSN 1874-6829 ; v. 6)

Includes bibliographical references and index.

1. Gesture. 2. Animal communication. 3. Primates. I. Pika, Simone. II. Liebal, Katja.

P117.D48 2012

156'.369--dc23 2011050944

ISBN 978 90 272 2848 2 (Hb ; alk. paper)

ISBN 978 90 272 7481 6 (Eb)

John Benjamins Publishing Co. · P.O. Box 36224 · 1020 ME Amsterdam · The Netherlands

John Benjamins North America · P.O. Box 27519 · Philadelphia PA 19118-0519 · USA

Elena Antinoro Pizzuto (1952–2011)

We devote this volume to Elena Antinoro Pizzuto whom we got to know with all her wit, joy and laughter during our first workshop on Gestural Communication in Non-human and Human Primates in Leipzig in 2004.

We are very sad about losing a colleague who was a pioneer in the linguistic study of signed languages and who contributed some of the classic studies in the field of language acquisition among deaf children. In addition to her concerted efforts to enable exchange and discussion between signed and spoken language linguists and her work on the development of writing systems for signed languages, Elena also studied the relation between signed language and gesture. As with all of her other endeavors, this work was truly state-of-the-art, insightful in ways rare to find. We will miss her dearly.

Table of contents

Acknowledgements

This book was inspired by a workshop on "Current developments in gesture research" held July 24th 2010 at the European University Viadrina in Frankfurt (Oder), Germany, at the 4th Conference of the International Society for Gesture Studies (ISGS).

We especially want to thank Jana Bressem, Silvia Ladewig, and Cornelia Müller, for their support in organizing this event and the Freie Universität of Berlin, who provided partial funding of this workshop.

Furthermore, we are grateful to the participants of this workshop as well as to Michael Arbib, Alan Costall, Catherine Crockford, Volker Deecke, Tobias Deschner, Judith Holler, Chris Knight, John Rae, Vasudevi Reddy, Wendy Sandler, Stuart G. Shanker, Björn Siemers, Chris Sinha, Michael Tomasello, Virginia Volterra, and Klaus Zuberbühler, who enabled a nurturing and fruitful discussion, resulting in lively debate and exchange to new and established avenues in gesture research.

Last but not least a big thank you to Cornelia Colsman, Julia Friedrich, Angelika Giessl and Carla Pritsch for their excellent support during the workshop and/or for helping with the index.

The editorial work for this book was supported by a Sofja Kovalevskaja Award of the Alexander von Humboldt Foundation to S.P.

About the contributors

Amanda E. Bania studies great ape cognition and behavior at the Smithsonian's National Zoological Park, Washington, USA. Her research interests include social cognition, tool use, causal understanding and reasoning abilities, and captive welfare.

Richard W. Byrne studies the evolution of cognition, particularly the origins of distinctively human characteristics, using evidence from species as diverse as great apes, elephants and domestic pigs. In 1987, with three colleagues, he set up the Scottish Primate Research Group, which now links 17 faculties and their research teams in an informal collaboration spanning five Scottish universities. He was awarded the British Psychology Society Book Award 1997 for his O.U.P. monograph "The Thinking Ape".

William D. Hopkins' main research focuses on the evolution of hemispheric specialization in primates. One central interest is the documentation of functional asymmetries in non-human primates, notably chimpanzees and other great ape species. This has included assessments of handedness, visual-half field studies and behavioral studies of facial expressions. As second focus is on neuroanatomical and functional asymmetries as measured through histological analyses done on post-mortem brains, in vivo magnetic resonance imaging and positron emission tomography.

Barbara J. King is Chancellor Professor of Anthropology at the College of William and Mary in Williamsburg, USA. Her books include The Dynamic Dance (2004) on great ape communication, and Being With Animals (2010) on animal-human bonding. Her Friday Animal Blog can be found at http://www.barbarajking.com.

David Leavens earned a B.S. in Anthropology from the University of California at Riverside, USA, in 1990, an M.S. in Anthropology from Southern Illinois University at Carbondale, USA, in 1993, and a Ph.D. in Biopsychology from the University of Georgia at Athens, USA, in 2001. Since 2000, he has been a member of the psychology faculty at the University of Sussex, UK. His central research interest is in how mental concepts, such as intentionality, are instantiated by particular bodies, in particular contexts, and how these patterns change with time and socio-ecological differences.

Katja Liebal is a Junior Professor for Evolutionary Psychology at the Freie Universität Berlin, Germany, Cluster *Languages of Emotion*. For her Ph.D., she worked on the gestural communication of gibbons and great apes at the Max Planck Institute for Evolutionary Anthropology in Leipzig, Germany, before she started a position as a lecturer at the University of Portsmouth, UK. Currently, her main research focus is on the multi-modal communication of emotions in primates.

Paula Marentette is a Professor of Psychology at the Augustana Campus of the University of Alberta, Canada. Her research focuses on gesture and its relationship to both spoken and signed languages in children and adults. She is particularly interested in the role of iconicity in gesture and sign.

Adrien Meguerditchian is studying gestural communication, handedness and brain specialization in non-human primates including baboons, chimpanzees and gorillas. Since his Ph.D. at the University Aix-Marseille in France, he works at the Yerkes National Primates Research Center in Atlanta and in a field work project on wild chimpanzees in Senegal, supported by the Fondation Fyssen. Adrien Meguerditchian is also part-time lecturer in Psychology, Ethology and Neuropsychology at the University Aix-Marseille in France.

Elena Nicoladis is an Associate Professor in the Department of Psychology at the University of Alberta, Canada. Her primary research interest is language and gestural development in bilingual and monolingual children. She is also interested in contextualizing language and gestural development within evolutionary, sociocultural and cognitive psychology.

Marcus Perlman is completing his Ph.D. in Cognitive Psychology at the University of California, Santa Cruz, USA, and currently works as a researcher at The Gorilla Foundation. His research examines the gestural and vocal communication of these great apes, with an emphasis on the role of iconicity in the evolution, development, and online use of language.

Simone Pika is the head of the Humboldt Research Group "Comparative Gestural Signalling" at the Max Planck Institute for Ornithology in Seewiesen, Germany. Her Ph.D. focused on gestural complexity of great apes, at the Max Planck Institute for Evolutionary Anthropology in Leipzig, Germany. She held research fellowships at the University of Alberta, Canada, the University of St. Andrews, Scotland and a post as Assistant Professor at the University of Manchester, UK.

Tim Racine is an Associate Professor in the Department of Psychology at Simon Fraser University, Burnaby, Canada. He is associated with the Developmental Psychology and Theory and Methods programs within his department, and his research interests include social cognitive development, the relation between evolution and development, and the philosophy of psychology.

His recent work has mostly focused on the origins of joint attention in human infants and non-human primates, and also the role of conceptual analysis and evolutionary explanation in psychology and related disciplines.

Steven J. Schapiro, Ph.D. is Chief of the Section of Primate Behavior and Environmental Enrichment at the Michale E. Keeling Center for Comparative Medicine and Research of The University of Texas MD Anderson Cancer Center in Bastrop, TX, USA. His primary research interest is behavioral management and he collaborates with numerous investigators studying many aspects of chimpanzee behavior, including lateralization and gestural communication.

Nicole M. Scott is currently a doctoral student in Cognitive Science at the University of Minnesota, USA. She received her B.S. in Biology from Iowa State University, USA, and MRes in Primatology from Roehampton University, UK. Her research interests concern comparative cognition of primates with a special focus on spatial knowledge and communication strategies.

Joanne Tanner received her doctorate in Evolutionary Psychology from the University of St. Andrews, Scotland, in 1998. She has continued to publish her observations of captive gorilla behavior as an independent researcher, in collaboration with colleagues. More information on Joanne's research can be found at http://www.gorillagestures.info/index.htm.

Sebastian Tempelmann earned his diploma in Biology in 2007 from the University of Leipzig, Germany, and is currently working on his Ph.D. at the Freie Universität Berlin, Germany. In addition, he works at the MPI for Evolutionary Anthropology in Leipzig, Germany, since 2004. His research interests concern the intentional communication of great apes and domestic dogs, with a special focus on the evolutionary roots of human referential communication.

Jan A.R.A.M. van Hooff (Arnhem, NL 1936) earned his Ph.D. at Universiteit Utrecht, the Netherlands, in 1971. He was a Professor of Ethology and Socio-ecology at Universiteit Utrecht, the Netherlands, from 1980 till 2001 with guest professorships at Universität Zürich, Switzerland, and at the École Normale Supérieur, Paris, France. His main research interests include primate social communication (e.g. the evolution of laughter and smiling) and comparative and evolutionary aspects of primate social organisation.

Sherman Wilcox is Professor of Linguistics at the University of New Mexico, USA. He is the author of several books, book chapters, and scholarly articles on signed language, specializing in the relation between gesture and language, the evolution of language, cognitive approaches to sign linguistics, and interpreting theory. His research includes studies of American Sign Language, Catalan Sign Language, and Italian Sign Language, and, most recently, Saudi Sign Language.

Introduction

Developments in primate gesture research

Simone Pika[1] & Katja Liebal[2]
[1]Max Planck Institute for Ornithology, Humboldt Research Group "Comparative Gestural Signalling", Eberhard-Gwinner Strasse, Geesehouse, 82319 Seewiesen, Germany / [2]Freie Universität Berlin, Berlin, Germany

Investigations into the gestural modality of humans have dramatically increased during the last two decades, focusing on a wide variety of topics ranging from the neural basis of gesture, over linguistic structures to cognitive functioning (e.g. Goldenberg & Strauss 2002; Goldin-Meadow 2003; Kendon 2001; McNeill 1992). These studies have also stirred new interest in uncovering the possible pre-cursors to language and the role gestural signals may have played during the dawn of human speech (e.g. Arbib, Liebal & Pika 2008; McNeill in press).

One useful avenue to unravel the mystery of language origins is the comparative approach, which pinpoints similarities and differences of communicative skills of non-human animals, especially our closest living relatives, the non-human primates. Although the vast majority of comparative researchers interested in language origins have neglected the gestural modality for a long time (for exceptions see; Corballis 2002; Hewes 1973), the first detailed descriptions of gestural signals of some great ape and monkey species have been provided by pioneers such as George Schaller (1964, 1965), Jane Goodall (Van Lawick-Goodall 1968a, 1968b), Jan van Hooff (1967, 1973), and Hans Kummer (1968). These ethograms contained many of the communicative gestures that have been observed in subsequent studies in the wild and in captivity and thus still represent very useful tools for current research. For instance in 2005, Pika and Mitani (2006) carried out a systematic study on a distinct gesture, the DIRECTED SCRATCH[1], which had first been described by Goodall (1968a: 264) studying the Gombe chimpanzees in Tanzania. She noted: "Indeed, deliberate scratching movements, during a grooming session, often served as signals since the partner normally responded by grooming the part scratched" (Goodall 1968a: 264). Pika and Mitani's study confirmed that this gesture truly is used as a communicative, referential signal to direct the attention

1. From here on gestures are depicted in small caps.

of recipients to distinct spots on the body to be groomed. In addition, they could show that DIRECTED SCRATCHES are mainly used between individuals with strong affiliative bonds such as "friends" and allies and acquire their meaning through dyadic shaping (Pika & Mitani 2009).

These first studies thus have and still do inspire several generations of researchers and also aided in developing new questions and research domains. One important domain in relation to language origins concerns the cognitive complexity underlying gestural usage; an area instigated by Frans Plooij (e.g. 1978, 1979). Studying the ontogeny of gestural signals in chimpanzees at Gombe, Tanzania, Plooij was the first to apply methods of Speech Acts Theory (Austin 1962) and parameters used in analyses of intentional behavior in human children (Bates, Camaioni & Volterra 1975; Piaget 1952). Plooij showed that chimpanzee infants' gestures are highly sophisticated communicative means (1) characterized by their flexible relation between means and ends (means-ends dissociation), and (2) used to attract and redirect attention. Means-ends dissociation implies that individuals are able to use (a) synonymous signals/gestures to achieve a certain outcome/goal, and (b) ambiguous gestures for different outcomes/goals. For instance the gestures TOUCHING and REACH OUT ARM are both used by chimpanzee infants to communicate to the mother to be picked up and thus carry the same message (Smith 1965)[2]. The gesture ARM RAISE on the other hand is used to solicit grooming but also to calm and appease an anxious conspecific (Plooij 1979). This gesture thus communicates and embodies different messages across contexts. Plooij (1978: 127) noted: "This indicates the ability to understand and to produce *new* meanings, and this suggest openness, which is one of the most characteristic design features of human language".

This cognitive approach to gestural signaling was continued and expanded by Michael Tomasello, Josep Call and their students. Their first studies focused on gestural development, flexibility and underlying learning mechanisms of chimpanzees in captivity (Tomasello, Call, Nagell, Olguin & Carpenter 1994; Tomasello et al. 1997; Tomasello, George, Kruger, Farrar & Evans 1985; Tomasello, Gust & Frost 1989), and were later extended by us to include the three other great ape species along with one species of small apes, respectively: bonobos (*Pan paniscus*; Pika 2007a; Pika, Liebal & Tomasello 2005), gorillas (*Gorilla gorilla*; Pika 2007b;

2. Smith (1965) introduced three distinct concepts to animal signaling: message, meaning, and context. The message of a signal may refer among other things to a generalized anxiety, an emotional state such as aggression or fear and does not necessarily imply the intention of a sender to communicate. The meaning is identified as the response selected by the recipient from all of the responses open to it, and the context refers to anything, which can be thought of as accompanying a signal.

Pika, Liebal & Tomasello 2003), orangutans (*Pongo pygmaeus*; Liebal 2007a; Liebal, Pika & Tomasello 2006) and siamangs (*Symphalangus syndactulus*; Liebal 2007b; Liebal, Pika & Tomasello 2004c).

This first comprehensive database on gestural signaling of all ape species provided evidence that ape's gestural competence is far more complex and sophisticated than their vocal abilities, showing that apes

- use open-ended, multifaceted gestural repertoires, including *species-distinctive* and *species-indistinctive* gestures, whose meaning and usage has to be learned (Pika et al. 2003; Call & Tomasello 2007);
- use gestures as flexibly produced intentional strategies, based on key characteristics utilized in studies of intentional communication in human children such as (a) recipient specificity, (b) persistence to the goal (e.g. repetition, exaggeration of a gesture or use of a different one until the goal has been achieved), (c) means-ends dissociation (see paragraph above) and (d) adjustment to audience effects such as (1) adaption of signal category to the attentional states of recipients, and (2) locomoting in the visual field of the recipient before producing a visual gesture (Liebal, Pika, Call & Tomasello 2004b; Call & Tomasello 2007); and
- develop group-specific traditions of gesture (Pika et al. 2003, 2005), implying underlying social learning processes involved.

Tomasello summarized the impact of these findings on scenarios of language evolution by noting: "In all, I personally do not see how anyone can doubt that ape gestures – in all of their flexibility and sensitivity to the attention of the other – and not ape vocalizations – in all of their inflexibility and ignoring of others – are the original font from which the richness and complexities of human communication and language have flowed"(Tomasello 2008: 55).

However, have we comparative psychologists and primatologists truly understood the phenomenon gesture? McNeill stated recently quite provocatively: "Despite the theory's name, the primatologist, neuroscientist, developmental psychologist, anthropologist, sign-language linguist, regular linguist, computer scientist, etc. proponents of gesture-first seemingly lack any serious acquaintance with gesture other than (it appears) its folk culture portrayals (so they do not recognize the key point of this book: that language is misconstrued if it is not seen as a unity of language and gesture)" (McNeill in press: 10).

We do not think that the lack of serious acquaintance with gesture and/or its relation with language is the main problem, but the lack of interdisciplinary exchange and the scientists' preponderance to use the same terms to talk about different things in addition to definitions of terms tailored to specific research objectives.

To address these problems, in 2004 we organized the first multidisciplinary workshop on "Gestural communication in non-human and human primates", resulting in a stimulating, interdisciplinary exchange and enterprise (Liebal, Müller & Pika 2007).

In addition, the last decade has seen a considerate increase of interest in gestural abilities of non-human primates. Although studies are still biased toward gestural behavior of apes in captivity (e.g. Cartmill & Byrne 2007; Pollick & De Waal 2007; Schneider, Call & Liebal 2010), there has been a considerate increase concerning gestural skills of apes in their natural environments (e.g. Genty, Breuer, Hobaiter & Byrne 2009; Hobaiter & Byrne 2011; Pika & Mitani 2006) as well as research on several monkey species (e.g. Hesler & Fischer 2007; Laidre 2008; Meguerditchian & Vauclair 2009; Smuts 2002).

In 2010, we thus initiated a second workshop at the European University Viadrina in Frankfurt (Oder), Germany to focus on "Current developments in primate gesture research". This time we were especially interested in enhancing and stirring debate(s) concerning:

- historical origins of gesture research;
- application and use of gesture definitions;
- importance and application of reliability measures;
- newest developments and trends in gesture research; and
- influences on theories of language evolution.

This book presents almost all of the papers of this workshop and some additional contributions, which originated and/or derived from lively exchange and discussion during and after it.

The book begins with an introductory reflection by Jan van Hooff, one of the leading pioneers of the field. Jan van Hooff started his influencing work over 50 years ago and provides insight into the historical roots of communication research with a special focus on facial expressions, gestures and vocalizations but discusses also the current state of the art in relation to language origins.

The second part of the book concerns empirical studies on gestural communication of humans and of non-human primates in captive settings and their natural environments. It starts with two chapters concerning the underlying learning mechanisms involved in the acquisition of gestures. This topic has recently stirred a lot of interest since scholars of one research group argue that apes do not learn their gestures (c.f. Call & Tomasello 2007) but simply rely on a genetically fixed, species-typical repertoire of unusually large size (Genty et al. 2009; Hobaiter & Byrne 2011). This hypothesis however does not solve the question why distinct gestures such as for instance the GROOMING-HAND-CLASP and LEAF-CLIPPING are used by some but not all chimpanzee communities in the wild (Whiten et al. 2001), and why some individuals use some gestures not utilized by any other group member (see picture 1, idiosyncratic gesture used by a single adult male (middle), Dolphi, over a period of five consecutive years at *Ngogo*, Kibale National Park, Uganda; ©Langergraber).

In their empirical study, Paula Marentette and Elena Nicoladis expand this debate by questioning whether children acquire some of their earliest gestures through a process of individual learning called *conventionalization* or *ontogenetic ritualization* (Clark 1978; Lock 1978; Tomasello 1996). They address this aspect by focusing on the use and development of five different gesture types in children between the ages of 6.5 and 11.5 months.

Marcus Perlman, Joanne Tanner and Barbara King illuminate the topic of non-human primate gesture acquisition from a different angle, by addressing the question whether gesture originates from instrumental action. To do so, they focus on the use of PUSHES, performed by a gorilla mother to direct her offspring around the enclosure.

The fifth chapter by Sebastian Tempelmann and Katja Liebal provides insight into the use of gesture sequences in orangutans with a special focus on the underlying linear structures and/or semantically referential information. This research area has received relatively little research attention and produced so far mainly conflicting results (e.g. Cartmill & Byrne 2010; Genty & Byrne 2010; Liebal et al. 2004a).

In the sixth chapter, Bill Hopkins and colleagues (Simone Pika, Katja Liebal, Amanda Bania, Adrien Meguerditchian, Molly Gardner & Steven Schapiro) present the first meta-analysis of data on handedness for manual gestures in all great ape species. They analyze species-level asymmetries in hand use and consistency of asymmetries across species to investigate whether left hemisphere lateralization for language in modern humans evolved from an existing lateralized system for gestural communication in the common ancestor of humans and apes.

The last two chapters of this subsection represent the cut set of empirical research and new trends in primate gestural communication. Mark Laidre focuses in the seventh chapter on form and function of visual gestures in a monkey species, the mandrill. He thus adds to the debate concerning gesture acquisition (see also Marentette & Nicoladis this volume; Perlman et al. this volume) and strengthens the view that in order to gain serious acquaintance with gesture we need to avoid "chimpocentristic" (Beck 1982) and "ape-centristic" views only.

Subsequently, Catherine Hobaiter and Richard Byrne present an empirical study on the use of gestures during consortships in chimpanzees at the *Sonso* community, Budongo Forest Reserve, Uganda. They suggest that gesture use is not always limited to evolutionary non-urgent functions (c.f. Tomasello & Zuberbühler 2002), and thus provide further evidence that the phenomenon gesture can only be understood in its full complexity if we study species living under natural selection pressures and in their natural environments (see also; Boesch 2007).

In the third part of the book, we intended to take the reader on a journey to explore and discover new trends and current debates in primate gesture research.

This third part starts with a chapter by Nicole Scott and Simone Pika, highlighting an ongoing debate on the application and use of gesture definitions (Liebal, Müller & Pika 2005). The authors show that definitions vary so tremendously between but also within scientific fields that significant comparisons of gestural complexity are rarely possible. In early ethological studies for instance, the term gesture is used without a definition (e.g. Van Lawick-Goodall 1968b), quite naturally implying that "everybody knows what it is" (James 1890). In our own work (e.g. Liebal et al. 2006; Pika 2008; Call & Tomasello 2007), we only consider those communicative signals as gestures, which are directed toward a recipient, are mechanically ineffective (which means, they are not designed to act as direct physical agents), and receive a voluntary response. Contrary, McNeill in his newest book defines a gesture as "an unwitting, non-goal-directed action orchestrated by speaker-created significances, having features of manifest expressiveness" (in press: 12), limiting the phenomenon gesture to the human species only. By combining McNeill's Gesture Continuum with Tinbergen's famous "four why's", Nicole Scott and Simone Pika therefore offer a new framework to gestural studies, intended to unite all gesture researchers within and across scientific fields.

The subsequent two chapters by Tim Racine and David Leavens focus on a contemporary debate over the phenomenon of POINTING, which has been long heralded as a human species-unique gesture. However, although POINTING has also been observed in non-human great ape species (e.g. Gardner & Gardner 1969; Inoue-Nakamura & Matsuzawa 1997; Miles 1990; Patterson 1978; Savage-Rumbaugh, McDonald, Sevcic, Hopkins & Rupert 1986; Woodruff & Premack 1979), some scholars argue that on a cognitive level human POINTING differs fundamentally from POINTING in apes (e.g. Butterworth 2003; Tomasello 2006; Tomasello, Carpenter & Liszkowski 2007). Contrary, other scholars suggest that POINTING is, to a relatively high degree, shared by all great ape species (e.g. Leavens & Racine 2009; Leavens, Racine & Hopkins 2009; Savage-Rumbaugh, Rumbaugh & McDonald 1985). To strengthen the latter viewpoint, Tim Racine provides an archaeology of the conceptual foundations of POINTING as exemplified through a critical analysis of one of the most influential theories in the field (Tomasello 2008; Tomasello, Carpenter, Call, Behne & Moll 2005; Tomasello 2009). Subsequently, David Leavens illustrates motivations and contexts of human POINTING to challenge the view that hypothetical motivations underlying POINTING differ fundamentally in humans and non-human great ape species.

The last chapter of this subsection by Ray Wilkinson, Ivan Leudar and Simone Pika represents another new trend in gesture research, – the necessity and fruitfulness of adapting and linking existing and developing new methodological tools. Ray Wilkinson and his colleagues combine methods of contemporary comparative research with a micro-analytic approach in the form of conversation analysis, previously used for human action only (ten Have 2007). They then apply this new tool to provide the first detailed insight in communicative sequences of chimpanzees during meat-sharing episodes at the *Ngogo* community, Kibale National Park, Uganda.

The final article by Sherman Wilcox leads us back to the uniqueness of human gesturing by focusing on the phenomenon of sign languages, which function without any use of speech (Klima & Bellugi 1979). This article pinpoints the developmental routes by which non-linguistic gestures become incorporated into the linguistic system of signed languages: one leading from gesture to word to grammatical morpheme, and a second leading from gesture to prosody to grammatical marker. Sherman Wilcox thus offers "new connections" and perspectives to comparative gestural approaches but also links back to the starting point of this book and Jan van Hooff's introductory reflections and the evolutionary origins of human language.

References

Arbib, Michael, Liebal, Katja and Pika, Simone. 2008. "Primate vocalization, gesture, and the evolution of human language." *Current Anthropology* 49(6): 1053–1076.

Austin, John L. 1962. *How to do things with words*. New York: Oxford University Press.

Bates, Elizabeth, Camaioni, Luigia and Volterra, Virginia. 1975. "The acquisition of performatives prior to speech." *Merrill-Palmer Quarterly* 21(3): 205–226.

Beck, Benjamin B. 1982. "Chimpocentrism: Bias in cognitive ethology." *Journal of Human Evolution* 11: 3–17.

Boesch, Christophe. 2007. "What makes us human (*Homo sapiens*)? The challenge of cognitive cross-species comparison." *Journal of Comparative Psychology* 3: 227–240.

Bruner, Jerome. 1981. "Intention in the structure of action and interaction." In *Advances in Infancy Research*, Lewis P. Lipsitt (ed.), Vol. 1, 41–56. New Jersey: Ablex, Norwood.

Butterworth, George. 2003. "Pointing is the royal road to language for babies." In *Pointing: Where language, culture, and cognition meet*, Sotaro Kita (ed.), 9–33. Mahwah, New York: Erlbaum.

Call, Josep and Tomasello, Michael. 2007. *The gestural communication of monkeys and apes*. Mahwah, New York: Lawrence Erlbaum Associates.

Cartmill, Erica A. and Byrne, Richard W. 2007. "Orangutans modify their gestural signaling according to their audience's comprehension." *Current Biology* 17(15): 1345–1348.

Cartmill, Erica A. and Byrne, Richard. W. 2010. "Semantics of primate gestures: Intentional meanings of orangutan gestures." *Animal Cognition* 13(6): 793–804.

Clark, Russell A. 1978. "The transition from action to gesture." In *Action, Gesture, and Symbol*, Andrew Lock (ed.), 231–260. New York: Academic Press.

Corballis, Michael C. 2002. *From Hand to Mouth, the Origins of Language*. Princeton, New Jersey: Princeton University Press.

Gardner, R. Allen and Gardner, Beatrice T. 1969. "Teaching sign language to a chimpanzee." *Science* 165: 664–672.

Genty, Emilie, Breuer, Thomas, Hobaiter, Catherine and Byrne, Richard. W. 2009. "Gestural communication of the gorilla (*Gorilla gorilla*): Repertoire, intentionality and possible origins." *Animal Cognition* 12(3): 527–546.

Genty, Emilie and Byrne, Richard. W. 2010. "Why do gorillas make sequences of gestures?" *Animal Cognition* 13(2): 287–301.

Goldenberg, Georg and Strauss, Stefan. 2002. "Hemisphere asymmetries for imitation of novel gestures." *Neurology* 59: 893–897.

Goldin-Meadow, Susan. 2003. *Hearing gesture: How our hands help us think*. Harvard: Harvard University Press.

Hesler, Nana and Fischer, Julia. 2007. "Gestural communication in barbary macaques (*Macaca sylvanus*): An overview." In *The Gestural Communication of Apes and Monkeys*, Josep Call and Michael Tomasello (eds), 159–196. Mahwah, NJ: Lawrence Erlbaum Associates.

Hewes, Gordon W. 1973. "Primate communication and the gestural origin of language." *Current Anthropology* 12(1–2): 5–24.

Hobaiter, Catherine and Byrne, Richard W. 2011. "The gestural repertoire of the wild chimpanzee." *Animal Cognition* 14(5): 747–767.

Inoue-Nakamura, Noriko and Matsuzawa, Tetsuro. 1997. "Development of stone tool use by wild chimpanzees (*Pan troglodytes*)." *Journal of Comparative Psychology* 111(2): 159–173.

James, William. 1890. *The principles of psychology*. New York: Holt.

Kendon, Adam. 2001. "Gesture as communication strategy." *Semiotica* 135: 191–209.

Klima, Edward S. and Bellugi, Ursula. 1979. *The Signs of Language*. Cambridge, Massachusetts: Harvard University Press.

Kummer, Hans. 1968. *Social organization of hamadryas baboons*. Chicago: University of Chicago Press.

Laidre, Mark. E. 2008. "Do captive mandrills invent new gestures?" *Animal Cognition* 11: 179–187.

Leavens, David A. and Racine, Timothy P. 2009. "Joint attention in apes and humans: Are humans unique?" *Journal of Consciousness Studies* 16: 240–267.

Leavens, David A., Racine, Timothy P. and Hopkins, William D. 2009. "The ontogeny and phylogeny of non-verbal deixis." In *The prehistory of language*, Rudie Botha and Chris Knight (eds), 142–165. Oxford: Oxford University Press.

Liebal, Katja. 2007a. "The gestural communication of orangutans." In *The gestural communication of monkeys and apes*, Josep Call and Michael Tomasello (eds), 69–98. Mahwah, New York: Lawrence Erlbaum Associates.

Liebal, Katja. 2007b. "The gestural communication of siamangs." In *The gestural communication of monkeys and apes*, Josep Call and Michael Tomasello (eds), 131–158. Mahwah, New York: Lawrence Erlbaum Associates.

Liebal, Katja, Call, Josep and Tomasello, Michael. 2004a. "Chimpanzee gesture sequences." *Primates* 64: 377–396.

Liebal, Katja, Müller, Cornelia and Pika, Simone. 2005. "Introduction: Gestural communication in nonhuman and human primates." *Gesture: Special issue* 5(1/2): 1–4.

Liebal, Katja, Müller, Cornelia and Pika, Simone. 2007. *Gestural communication in nonhuman and human primates*. Amsterdam/Philadelphia: John Benjamins Publishing Company.

Liebal, Katja, Pika, Simone, Call, Josep and Tomasello, Michael. 2004b. "Great ape communicators move in front of recipients before producing visual gestures." *Interaction Studies* 5(2): 199–219.

Liebal, Katja, Pika, Simone and Tomasello, Michael. 2004c. "Social communication in siamangs (*Symphalangus Syndactulus*): Use of gestures and facial expression." *Primates* 45(2): 41–57.

Liebal, Katja, Pika, Simone and Tomasello, Michael. 2006. "Gestural communication of orangutans (*Pongo pygmaeus*)." *Gesture* 6(1): 1–38.

Lock, Andrew. 1978. *Action, Gesture and Symbol: The emergence of language*. New York: Academic Press.

Marentette, Paula and Nicoladis, Elena. 2012. "Does ontogenetic ritualization explain early communicative gestures in human infants?" In *Developments in Primate Gesture Research*, Simone Pika and Katja Liebal (eds), 33–53. Amsterdam/Philadelphia: John Benjamins Publishing Company.

McNeill, David. 1992. *Hand and Mind*. Chicago: The University of Chicago Press.

McNeill, David. in press. *How Language Began: Gesture and Speech in Human Evolution*. Cambridge: Cambridge University Press.

Meguerditchian, Adrian and Vauclair, Jacques. 2009. "Contrast of hand preferences between communicative gestures and non-communicative actions in baboons: Implications for the origins of hemispheric specialization for language." *Brain and Language* 108(3): 167–174.

Miles, H. Lyn. 1990. "The cognitive foundations for reference in a signing orangutan." In *Language and Intelligence in Monkeys and Apes*, Sue T. Parker and Kathleen R. Gibson (eds), 511–539. Cambridge: Cambridge University Press.

Patterson, Francine. 1978. "Conversations with a gorilla." *National Geographic* 134(4): 438–465.

Perlman, Marcus, Tanner Joanne E. and Barbara J. King. 2012. "A mother gorilla's variable use of touch to guide her infant: Insights into iconicity and the relationship between gesture and action." In *Developments in Primate Gesture Research*, Simone Pika and Katja Liebal (eds), 55–71. Amsterdam/Philadelphia: John Benjamins Publishing Company.

Piaget, Jean. 1952. *The Origins of Intelligence in Children*. New York: Norton.

Pika, Simone. 2007a. "The gestural communication of bonobos." In *The gestural communication of monkeys and apes*, Josep Call and Michael Tomasello (eds), 41–67. Mahwah, New York: Lawrence Erlbaum Associates.

Pika, Simone. 2007b. "The gestural communication of gorillas." In *The gestural communication of monkeys and apes*, Josep Call and Michael Tomasello (eds), 99–130. Mahwah, New York: Lawrence Erlbaum Associates.

Pika, Simone. 2008. "What is the nature of the gestural communication of great apes?" In *The Shared Mind*, Jordan Zlatev, Timothy P. Racine, Chris Sinha and Esa Itkonen (eds), 165–186. Amsterdam/Philadelphia, the Netherlands: John Benjamins Publishing Company.

Pika, Simone, Liebal, Katja and Tomasello, Michael. 2003. "Gestural communication in young gorillas (*Gorilla gorilla*): Gestural repertoire, learning and use." *American Journal of Primatology* 60(3): 95–111.

Pika, Simone, Liebal, Katja and Tomasello, Michael. 2005. "Gestural communication in subadult bonobos (*Pan paniscus*): Gestural repertoire and use." *American Journal of Primatology* 65(1): 39–51.

Pika, Simone and Mitani, John C. 2006. "Referential gesturing in wild chimpanzees (*Pan troglodytes*)." *Current Biology* 16(6): 191–192.

Pika, Simone and Mitani, John C. 2009. "The *directed scratch*: Evidence for a referential gesture in chimpanzees?" In *The Prehistory of Language*, Rudie Botha and Chris Knight (eds), 166–180. Oxford: Oxford University Press.

Plooij, Frans X. 1978. "Some basic traits of language in wild chimpanzees?" In *Action, Gesture and Symbol*, Andrew Lock (ed.), 111–131. London: Academic Press.

Plooij, Frans X. 1979. "How wild chimpanzee babies trigger the onset of mother-infant play." In *Before Speech*, Margaret Bullowa (ed.), 223–243. Cambridge: University Press.

Pollick, Amy S. and De Waal, Frans B. M. 2007. "Ape gestures and language evolution." *Proceedings of the National Academy of Sciences of the United States of America* 104: 8184–8189.

Savage-Rumbaugh, Sue E., McDonald, Kelly, Sevcic, Rose A., Hopkins, William D. and Rupert, Elizabeth. 1986. "Spontaneous symbol acquisition and communicative use by pygmy chimpanzees (*Pan paniscus*)." *Journal of Experimental Psychology: General* 115: 211–235.

Savage-Rumbaugh, Sue E., Rumbaugh, Duane. M. and McDonald, Kelly. 1985. "Language learning in two species of apes." *Neurosciences and Biobehavioral Review* 9: 653–656.

Schaller, George. B. 1964. *The year of the gorilla*. Chicago: University of Chicago Press.

Schaller, George. B. 1965. "The behaviour of the mountain gorilla." In *Primate Behaviour*, Irven de Vore (ed.), 324–367. New York: Holt, Rinehart and Winston.

Schneider, Christel, Call, Josep and Liebal, Katja. 2010. "Do bonobos say NO by shaking their head?" *Primates* 51(3): 199–202.

Smith, W. John. 1965. "Message, meaning, and context in ethology." *American Naturalist* 908: 405–409.

Smuts, Barbara. 2002. "Gestural communication in olive baboons and domestic dogs." In *The cognitive animal: Empirical and theoretical perspectives on animal cognition*, Marc Bekoff, Colin S. Allen and Gordon Burghardt (eds), 301–306. Cambridge, MA: MIT Press.
ten Have, Paul. 2007. *Doing Conversation Analysis*. London: Sage.
Tomasello, Michael. 1996. "Do apes ape?" In *Social Learning in Animals: The Roots of Culture*, Cecilia M. Heyes and Bennett Galef Jr. (eds), 319–346. San Diego: Academic Press, Inc.
Tomasello, Michael. 2006. "Why don't apes point?" In *Roots of Human Sociality: Culture, Cognition and Interaction*, Nicholas J. Enfield and Stephen C. Levinson (eds), 506–524. Oxford: Berg.
Tomasello, Michael. 2008. *Origins of Human Communication*. Cambridge, Massachusetts: MIT Press.
Tomasello, Michael, Call, Josep, Nagell, Katherine, Olguin, Raquel and Carpenter, Malinda. 1994. "The learning and use of gestural signals by young chimpanzees: A trans-generational study." *Primates* 35(2): 137–154.
Tomasello, Michael, Call, Josep, Warren, Jennifer, Frost, Thomas G., Carpenter, Malinda & Nagell, Katherine. 1997. "The ontogeny of chimpanzee gestural signals." In *Evolution of Communication*, Sherman Wilcox, Barbara King and Luc Steels (eds), 224–259. Amsterdam/Philadelphia: John Benjamins Publishing Company.
Tomasello, Michael, Carpenter, Malinda, Call, Josep, Behne, Tanya and Moll, Henrike. 2005. "Understanding and sharing intentions: The origins of cultural cognition." *Behavioral and Brain Sciences* 28: 1–17.
Tomasello, Michael, Carpenter, Malinda and Liszkowski, Ulf. 2007. "A new look at infant pointing." *Child Development* 78(3): 705–722.
Tomasello, Michael, George, Barbara L., Kruger, Ann C., Farrar, Michael J. and Evans, Andrea. 1985. "The development of gestural communication in young chimpanzees." *Journal of Human Evolution* 14: 175–186.
Tomasello, Michael, Gust, Deborah and Frost, G. Thomas. 1989. "A longitudinal investigation of gestural communication in young chimpanzees." *Primates* 30(1): 35–50.
Tomasello, Michael and Zuberbühler, Klaus. 2002. "Primate vocal and gestural communication." In *The cognitive animal: Empirical and theoretical perspectives on animal cognition*, Marc Bekoff, Colin S. Allen and Gordon Burghardt (eds), 293–299. Cambridge: MIT Press.
Tomasello, Michael. 2009. *Why We Cooperate*. MA: MIT Press.
Van Hooff, Jan A. R. A. M. 1967. "The facial displays of the catarrhine monkeys and apes." In *Primate Ethology*, Desmond Morris (ed.), 7–68. London: Weidenfeld and Nicolson.
Van Hooff, Jan A. R. A. M. 1973. "A structural analysis of the social behaviour of a semi-captive group of chimpanzees." In *Social Communication and Movement, Studies of Interaction and Expression in Man and Chimpanzee*, Mario von Cranach and Ian Vine (eds), 75–162. London: Academic Press.
Van Lawick-Goodall, Jane. 1968a. "The behavior of free-ranging chimpanzees in the Gombe Stream Reserve." *Animal Behaviour Monographs* 1(3): 161–311.
Van Lawick-Goodall, Jane. 1968b. "A preliminary report on expressive movements and communication in the Gombe stream chimpanzees." In *Primates. Studies in Adaptation and Variability*, P. C. Jay (ed.), 313–374. New York: Holt, Rinehart, and Winston.
Whiten, Andrew, Goodall, Jane, McGrew, William C., Nishida, Toshisada, Reynolds, Vernon, Sugiyama, Yukimaru, et al. 2001. "Charting cultural variation in chimpanzees." *Behaviour* 138: 1489–1525.
Woodruff, Guy and Premack, David. 1979. "Intentional communication in the chimpanzee: The development of deception." *Cognition* 7: 333–352.

Facial-vocal displays, gestures and language

Jan A.R.A.M. van Hooff
Universiteit Utrecht, The Netherlands

Animals possess nonverbal displays consisting of chemical signals, postures, gestures, calls, and, in mammals, of facial-vocal expressions, essentially reflecting intentions and attitudes of the sender. Humans, in addition, possess a verbal language with an 'unlimited' repertoire of referential and representational symbols, syntactically combined into meaningful complexes. Because of the vital role of speech the emancipation of vocal utterances in primates is seen as the beginning of linguistic evolution. However, in humans gesturing is closely associated with linguistic expression, favouring a gestural origin. Here I compare features of gesturing and facial-vocal expression in our primate relatives, e.g. referentiality and intentionality, contrasting these with linguistic performance. The conclusion is that a choice between one of the former as the main precursor of language cannot as yet be made.

1. The study of communication in primates

The invitation to write an introductory reflection on the relationships between facial-vocal displays, gestures and language took me back just over 50 years, when, in the fall of 1960, I started my study of primate facial-vocal expressions in the London Zoo, as a master student of Niko Tinbergen. It seemed, I entered a largely untrodden field. Well, at least not lately trodden, because there was, of course, Darwin's pioneering work "On the expression of the emotions in man and the animals", published nearly hundred years earlier (Darwin 1872). Darwin's interest clearly was an evolutionary one: comparing the expressions of the mood in related species might provide evidence that also the mind must be understood as a product of evolution.

The defining characteristic of the human mind is language, and its most noticeable channel is speech. So considerations about its evolutionary origin naturally zoomed in on the vocal utterances of non-human primates. But then the finding that the great apes had a repertoire of flexible and volitional gestures led to the hypothesis, that the emancipation of these gestures into referential symbols has been at the basis of the development of language in our species. Lately evidence is accumulating for a referential function of primate vocalizations, and the 'vocal hypothesis' is revived (e.g. Zuberbühler et al. 2009). This merits a comparison of relevant characteristics of facial-vocal displays, gestures and language.

The evolution of communicative behaviors has always fascinated early ethologists. Accordingly their focus was on species-typical patterns, where a strong genetic component implies that these have been shaped by evolutionary selection. Hence the search was also for between-species similarities in species-typical displays that would reveal evolutionary homologies. That enterprise has its pitfalls. For example, the silent bared-teeth display of *Mandrillus* has been considered homologous with the human expression of rage. At first sight, it strongly resembles our snarl face (see Figure 1) and the facial expression is

Figure 1. A mistaken homology: The "friendly" horizontal bared-teeth face is regarded homologous with the human expression of rage. From Eibl-Eibesfeldt and Sütterlin, 1992

accompanied by vehement horizontal head shaking, "as though tearing at an opponent"(Andrew 1963:65). In fact the head movement is the exact opposite of the abrupt vertical head nod, which is part of the threat signal of this species. Thus it is a magnificent example of Darwin's 'principle of antithesis', namely that displays with contrasting messages have become ritualized and exaggerated in opposite directions. Because that is what the display is: the opposite of aggressivity, an indication of non-hostility, of affiliation (van Hooff 1965; Bout & Thierry 2005). Clearly inferences about evolutionary relationships should rest on broad comparative structural, motivational and functional analyses, such as those done, in the case of the primate silent bared-teeth displays, for a number of macaque species by Preuschoft (1995).

Given the interest in interspecific variation, intraspecific variation was seen rather as disturbing noise. In this respect mammals seemed to differ from other taxa where species-typical 'fixed action patterns' could be distinguished comparatively easily. Yet Lorenz (1943) noted that the greater variability in mammals does not pre-empt attempts to classify. Thus he showed for dogs that the seemingly endless variation in facial expression could be understood as the variable activation of two conflicting behavioral tendencies. This accorded well with the way in which the displays of birds and fish are considered to result from conflicting behavioral tendencies, e.g. the zig-zag dance of the stickleback (Tinbergen 1954).

Also in the facial-vocal displays of non-human primates species-typical patterns could be detected (van Hooff 1962, 1967; Andrew 1963). Similarly there is now consensus that also in our species there are a number of species-typical, universally present facial-vocal displays (e.g. Ekman & Friesen 1975). These have been related to the categories of our basic emotions that have emerged from numerous discussions between psychologists (see Ortony & Turner 1990). About seven categories of emotion cropped up consistently, namely the categories "interest", "surprise and astonishment", "anger and rage", "fear and terror", "sadness, anguish and distress", "joy, pleasure and happiness" and "disgust and contempt", each characterised by a more or less specific facial expression (e.g. Plutchik 1980; Frijda 1987). Five of these can be equated with categories of ritualized facial expression that are generally present in non-human primates (van Hooff 1976).

However, for two categories of facial expression that have an emotional correlate in man, there seem to be no ritualized counterparts in non-human primates. These are the expressions of "surprise and astonishment" and "disgust and contempt". This may well be related to a noteworthy difference between us and the other primates. Whereas their displays primarily express a motivational attitude towards a companion, we use our facial-vocal displays often in a declarative referential way. They almost continuously underline, modulate and illustrate our verbal comments. Sometimes these signals are emphatic, sometimes fleeting, such as the momentary raising of the eyebrows in a doubting remark, or the compressing and bulging of the lips in an assertive comment. Such declarative functions are especially evident with respect to the ritualized expressions of surprise/astonishment and disgust/contempt, which are lacking in our relatives.

This brings us to the question where we find the origins of that unique characteristic of our own species, the faculty of language and, especially, its manifestation in the form of speech? Has it been emancipated from primate vocal utterances? Or should we seek its origin elsewhere, namely in primate gesturing?

Compared with primate facial-vocal displays the postural and dynamic variation in the gestures has seemed to be large, at least in the apes. This made it less easy to distinguish fixed action patterns, so I noted, when describing hand and arm postures and movements during my ethological studies of chimpanzee behavior in the late sixties (van Hooff 1973, see Figure 2). The difference in flexible applicability between facial-vocal behaviors and

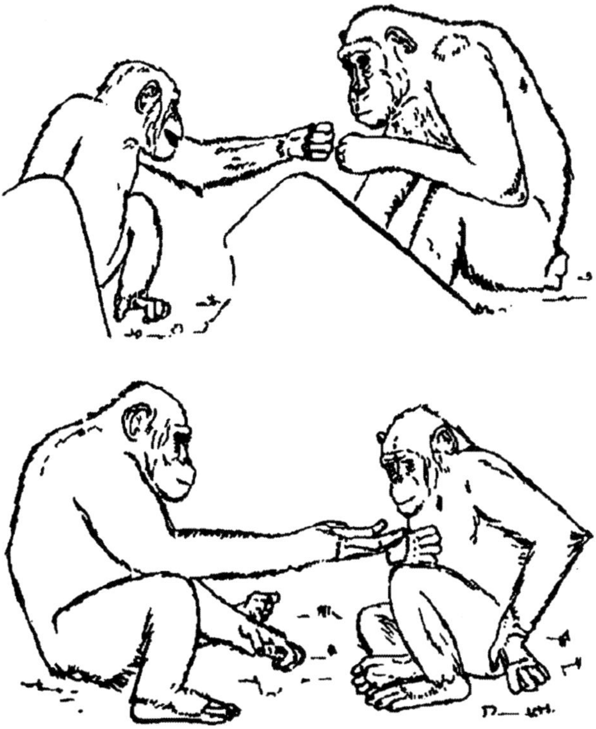

Figure 2. Two common versions of 'hold out hand'
top: Hand vertical, wrist and fingers bent; the actor shows also 'bared-teeth yelp'
bottom: Hand horizontal, wrist and fingers fully stretched

gestures became particularly apparent with the endeavours to teach chimpanzees a verbal language. An early attempt to teach a chimpanzee to speak yielded very poor results (Hayes & Hayes 1952). However, it has appeared possible to teach great apes a variety of symbolic gestures (e.g. Gardner & Gardner 1969; Fouts 1973), as well as arbitrary lexigrams (e.g. Premack 1971; Rumbaugh 1977; Matsuzawa 1985, 2003). Similar results have been obtained with cetaceans and pinnipeds (e.g. Schusterman & Gisiner 1988), and remarkably – using the vocal channel – with birds (Pepperberg 2000).

Teaching these symbols required immersion in human-animal interaction and intensive training. With respect to the latter there is a striking contrast with the ease and spontaneity with which very young children start babbling, not only vocally, but also manually (e.g. Petitto et al. 2004) and acquire the speech characteristics of a language (e.g. Kuhl 2000). Nevertheless the ape language studies were a remarkable achievement, even though the surprising initial results gave rise to assumptions about the competences of these apes that ran too high (Terrace 1983; Rivas 2005). The studies have strengthened suppositions that the evolution of speech has arisen in the wake of an evolutionary development of increasing use of gestures as voluntary and intentionally used symbols (e.g. review by Hewes 1973). This would have happened in concomitance with adaptations of both the neuronal and anatomical substrates enabling differentiated vocalization (Lieberman et al. 1969), when the production of vocalizations became, so to speak, liberated from its emotional constraints. Recent neurobiological and neuropsychological findings (e.g. Rizzolatti & Arbib 1998) are advanced as arguments in support of an origin in manual gesturing (e.g. reviews by Rizzolatti & Sinigaglia 2008; Corballis 2009).

In the following, I will compare the different characteristics in which facial-vocal displays and language differ. I will consider to what extent gestures share these characteristics and whether that makes them a better candidate as the starting point for language development.

Table 1. Supposed differences between facial-vocal displays, gestures/postures and language

Facial-vocal displays	Gestures/postures	Language
[1] Species-typical, "genetic"	Ontogenetic Ritualization	Learned culturally
[2] Signals of "attitude", analog	Iconic signals, analog	Symbols, arbitrary, digital
[3] Closed, limited repertoire	Greater flexibility	Open, unlimited repertoire
[4] Emotionally regulated (only?)	More intentional, voluntary	Intentional, voluntary
[5] Referential: very little	Perhaps indicating referents	Essentially referential
[6] Functional expression	Imperative, instrumental	Instrumental + Conversational
[7] Combinations (but syntax?)	Syntax? No evidence	Syntax essential, generative grammar

1.1 Species-typical, ontogenetically ritualized or learned culturally

Human language competence clearly is a species-specific and, therefore, a phylogenetically adapted attribute. It remains controversial, however, how specific the genetic programming is of certain dispositions that underlie language acquisition (e.g. Deacon 1998; Tomasello 2003; versus Chomsky 1982; on 'innate, universal grammar'). Clearly the actual verbal symbols or words are culturally acquired. By contrast the facial-vocal displays of primates and many other species show taxonomic specificity and intra-species universality, both in their ethographic and their motivational and functional aspects. This also holds for human facial-vocal behavior, although there is cultural variation in the display rules (Ekman & Friesen 1975). Thus, primate gestures seem to be somewhere in between language and facial-vocal displays. Some natural gestures can be regarded as an intention movement or as the utterance of a conflict of different behavioral tendencies. Mutual feedbacks in the interactions between conspecifics would then have led to a process of, as Tomasello (1996) has termed it, ontogenetic ritualization (see also Marentette & Nicoladis this volume; Perlman et al. this volume; Pika & Liebal this volume). For an example see the chimpanzee "stretch over" (van Hooff 1973), performed by a male as an invitation to a female to crouch-present in front of him (Figure 2b). Such ontogenetic ritualization is analogous to the evolutionary ritualization that has been studied by ethologists in connection with the evolution of species-typical displays, especially in birds (Tinbergen 1954).

Figure 3. *top:* A male stretches one hand over the back of a female while preparing to mount
bottom: The 'stretch over' performed to a female from a distance
Drawings from photographs by J.v.Hooff. From van Hooff (1973)

Pollick and de Waal (2007) emphasise the significance of tradition for the variation of gestures. For instance two adjoining chimpanzee populations in East Africa were found to have an idiosyncratic form of manual contacts during grooming, not found in other populations; in one population this was a hand-hand clasp; in the other a hand-wrist clasp (McGrew et al. 2001).

Not all gestures are shaped by ontogenetic ritualization though. Genty et al. (2009) found no evidence for gorilla gestures; they claim that they are species-typical and genetically specified, as for instance the chest beating of gorillas (however, see Pika et al. 2003 for a different distinction).

1.2 Iconic analog signals versus arbitrary digital symbols

Facial-vocal displays can be regarded as ritualized incipient movements or manifestations of behavioral conflicts (Darwin 1872; Andrew 1963; van Hooff 1967). They are iconic representations and they are of an analog nature, i.e. graded signals. The same applies to the gestures occurring in the natural social interaction of monkeys and apes. In this respect gestures resemble the facial-vocal displays. In contrast, human verbal symbols are of a digital nature and as a rule completely arbitrary. Yet the contrast with primate calls is not as strong as supposed earlier. Recent studies of vocalizations of non-human primates have revealed the existence of referential calls (see [5]) that are arbitrary and cannot be explained as iconic signals. Even though they appear to be species-typical these vocalizations resemble verbal symbols much more than any natural gestures described so far.

However, ape "language" studies show that the potential for learning to use arbitrary gestures, and for that matter, also visual patterns (lexigrams), as referential symbols is considerable, at least in human-enculturated individuals.

1.3 Closed, limited repertoire versus open, unlimited repertoire

The human verbal repertoire is open and unlimited in principle; new words can be added as the need arises. By contrast the number of facial-vocal displays is limited within species-specific bounds. To a large extent this also applies to the repertoire of gestures as used by apes in their natural interactions. How many gestures are being described for a primate species depends on whether the investigator is a "splitter" or a "lumper". Thus Pika et al. (2003) mention 33 gestures and Tanner (1998) 30 for the gorilla, whereas Genty et al. (2009) distinguish 102 gestures. It makes a difference, for example, whether one distinguishes the 'stretch arm, open hand with palm up' as a behavioral element or whether one distinguishes the same motor pattern as a 'begging for food', 'begging for support' or 'begging for reconciliation' signal.

The latter authors report that the intra-community and the inter-community variations are about the same, emphasizing that there is a limited species-typical repertoire. In this respect there may, however, be differences between ape species. Thus populations of the genus *Pan* differed in the use of gestures, but not in the use of facial-vocal displays (Pollick & de Waal 2007). The differences were greater, moreover, in the bonobo than in the chimpanzee.

1.4 Emotionally regulated versus intentional and voluntary

An important difference between facial-vocal displays and gestures/postures concerns the degree to which these are under voluntary control and can be intentionally applied. This difference became obvious early on when attempts to teach chimpanzees to use vocal utterances as the carriers of verbal symbols appeared to have little success (Hayes & Hayes 1952). By contrast attempts to use gestures and script symbols have been very successful (e.g. Gardner & Gardner 1969; Fouts 1975; Rumbaugh 1977; Matsuzawa 1985). This indicates that the failure to teach these apes verbal symbols based on speech is not due to a lack of cognitive competence, i.e. the capacity to store and associate the relevant information, but that the facial-vocal utterances are tied to and constrained within the framework of emotional expression. A telling example of the difference between facial expression and gesture in voluntary command has been given by de Waal (1982): A chimpanzee was observed to hide her fearful bared-teeth display by covering her face with her hand.

This emotional constraint had to become loosened in our ancestors for speech to become an instrument of language. Voluntary control is easier with 'low intensity' vocalizations where the emotional involvement is less, such as the grunts and facial movements described for orangutans by Cartmill and Byrne (2010). The mentioned constraint seems to be a general characteristic of mammals. But it is certainly not absolute, as is illustrated by an observation made by Rüppell (1969). A female arctic fox uttered the species-typical alarm call to drive away her cubs from food that she had brought to them, but that was no longer accessible to herself, because the youngsters crowded upon it. Rather than driving them off aggressively, she had learned to use the call purposively, to the same desired effect (*cf.* the manipulative barking that can be observed also in dogs in interaction with humans). Notwithstanding such examples, there is a strong contrast between mammals and birds in the freedom of the use, and the imitative and creative ability in the production of vocal utterances that is shown by some bird taxa, notably the legendary mynas and parrots (e.g. Pepperberg 2000).

It is almost by definition that voluntary movements can be employed intentionally. But is the reverse also true; is a so-called 'instinctive' and species-typical behavior always unintentional? Much depends on how intentionality is defined and has been operationalised. It makes sense, first of all, to distinguish between the ultimate and proximate functionality of a biological process. On the one hand there is the orientation towards an ultimate adaptive effect that has been brought about by natural selection; it fine-tunes a process so that this adaptive effect is obtained most effectively and efficiently. On the other hand there is the proximate, the here-and-now regulation of the process. The ultimate effect often is not a factor in the proximate regulation. For example, when, in response to a threatening stimulus, the bittern freezes in its typical vertical position with its neck stretched and bill 'pointing' upwards, this makes it blend in better with the reeds. One may say that the behavior is intended, or functions to mislead predators, but the consequences of neither the blending nor the misleading are feedback factors that regulate this almost reflex-like response. The bittern has no intention to mislead. Similarly there is no reason to suppose that waggle-dancing honeybees intend to inform fellow bees, although that is the function of this 'pointing' from an evolutionary perspective.

But now take the classic experiment by Lorenz and Tinbergen (1938) on egg retrieval by blue geese. The authors made the significant distinction between the motor and the taxis component of a behavior. The latter is the feedback-regulated steering of the rolling egg so that it ends up in the nest cup, i.e. what the bird "aims at". Ethologists and comparative psychologists have long since recognised such behavior as 'purposive'. Another qualification is 'intentional', but this term has a mentalistic flavour and may be taken to imply 'conscious control'. Because even in humans intentionality does not always involve cognitive awareness, this cannot be a defining feature of intentionality. Ambiguity might be avoided by using the generally applicable and well operationalised term 'purposive' rather than 'intentional' in cases where the effects of a behavior do feedback on the functional adjustment of the process and where particular effects are thus identifiable as what the animal is trying to achieve. When, in addition to goal directed adjustment, there is also flexibility in that different means can be chosen to reach the goal, and the same means can be applied to different ends, then we more readily assume that there is intelligent purposiveness.

Figure 4 illustrates the relevance of the above. The species-typical facial-vocal display of the chimpanzee on the left was clearly a highly emotional expression of utter frustration, but it was also regulated in response to the effects it had on the behavior of the companion. It was clearly aimed at being allowed access to the banana. No doubt, also 'highly emotional facial-vocal displays can be purposively modulated by audience feedbacks (see Slocombe & Zuberbühler (2007) for an example of adaptive modulation of recruitment calls in chimpanzees).

Figure 4. "Don't refuse me a banana!" (Photo J.v. Hooff)

In short: both facial-vocal displays and gestures may be used purposively, whether they are instinctual or not.

1.5 Displays of attitude or referential symbols

Attempts to teach animals such as apes, parrots, pinnipeds or cetaceans to use gestures as referential symbols have been successful. In animals living under natural conditions referential vocalizations have been found. In 1980, Seyfarth et al. for instance reported about three types of alarm call in vervet monkeys that each indicate a specific type of predator, namely a leopard, an eagle or a snake. Lately systems of functional referential vocalizations have been found in other guenons, namely in stub-nosed monkeys (Arnold et al. 2008), in Campbell's monkeys (Ouattara et al. 2009), in chimpanzees (Hallberg et al. 2003) and also in birds (Evans & Evans 2007). The calls appear to belong to a small, closed and species-typical repertoire. Therefore, a resemblance to words in human language is still very superficial, and they have been considered to be of little relevance for a theory of human linguistic evolution.

Till recently there was no evidence of a referential use of vocalizations in apes. This led to the opinion that the crucial trajectory towards the evolution of a capacity for language in the hominid lineage has gone primarily via the emancipation of the referential use of gestures. However, recently Slocombe and Zuberbühler (2005) have shown that in the so-called 'rough grunts' of chimpanzees different grunt types can be distinguished. These are short, low intensity vocalizations, which are uttered in the context of feeding. When in a play-back experiment a particular grunt type was presented to a subject it stimulated the animal to search for a particular food type. Slocombe and Zuberbühler (2005) thus argue that the grunts are meaningful referents for conspecifics in their natural communication and propose a revaluation of the vocal hypothesis of language evolution vis-à-vis the gesture hypothesis (Slocombe & Zuberbühler 2010). A crucial argument here would be how the differentiation arises ontogenetically. Are the grunt differences genetically hard-wired as is likely for the guenon alarm calls or do they result from some or other form of ontogenetic and cultural ritualization? Comparative developmental studies in different (captive) populations should give the answers.

There is little evidence for referential gestures and body movements in natural circumstances. The nearest to it are various forms of 'pointing', indicating directions for others to follow, such as, in birds, the behavior of the honey guide (very appropriately called *Indicator indicator;* Short & Horne 2002), in bees the circle and waggle dances to indicate the direction in which and the distance at which a food source can be found (von Frisch 1965).

'Pointing' has drawn particular interest as a form of purposive referential signaling. In humans it occurs in a variety of forms, such as the 'pointing' with the index (!) finger, and it appears already in the pre-verbal stage. Chimpanzees can learn to use it in interactions with humans (Menzel 2005; Leavens & Hopkins 2007). So far the evidence for referential 'pointing' in feral animals is limited to a few examples; for instance of chimpanzee males of the *Ngogo* community, Uganda that 'point' to parts of their bodies to be groomed (Pika & Mitani 2006). However, other forms of 'pointing' do occur; see also Leavens (this volume).

A conspicuous form of facial-postural 'pointing', explicitly purposive and occurring, moreover, in a triadic context, has been studied in long-tailed macaques (de Waal et al. 1976). During conflicts individuals may solicit coalition partners to join them against a particular opponent by 'showlooking', an abrupt alternation between looking at the partner and looking at the opponent, with the neck stretched forward, lifted eyebrows and with protruded lips.

A remarkable form of pointing-like behavior exists in the hamadryas baboon (Stolba 1979). Before the one-male groups that belong to the same clan disperse in the early morning they sit together. Then some of the males may make a short walk, each in a different direction. When other males back one of the 'proposed' directions by 'going along' for a short distance, the groups part and go off to forage, each in a different direction. Around noon they reassemble at a drinking pool. Stolba deduced that what he had seen in the morning was a kind of 'voting' process in which the males decide about a meeting destination, proposed by 'pointing' (see also Kummer 1996).

This is more than the prompting of an action the signer wants another to do by performing a behavior that depicts the desired action iconically, as described by Tanner and Byrne (1996) for a gorilla. Sure, the other baboons went along in the same direction for a moment, but then they often went their own ways. What makes this case remarkable is that the message is not about the actual direction to move, but about the eventual destination, which is out of sight at the time of the 'vote and decide' process. The harems reach this destination eventually, even after they have meandered in different directions. Thus it seems to differ from the usual problem that groups have, namely to move all *together* in spite of diverging individual interests (Conradt & Roper 2005). We must conclude that the 'pointing' evokes an imagination in the onlookers and that they thus share their imaginations. This is the hallmark of the linguistic exchange of information and cultural cognition in our species (cf. Tomasello et al. 2005). Still, we do not know whether the behavior of the individuals had any communicative intent. The continuous mutual checking by actors and audience would suggest that it is purposively regulated and thus is unlike the non-purposive 'pointing' by the waggle-dancing bees, discussed under [4]).

1.6 Imperative and instrumental versus attention sharing and narrative conversation

A significant difference between human speech and human gesturing on the one hand and primate gesturing on the other hand became visible as a result of both of the ape sign language studies and of studies comparing the 'pointing' observed in chimpanzees with that of prelingual children and its ontogeny (Tomasello et al. 2007; Liebal et al. 2009). Whereas chimpanzees can learn to 'point' in a context where it is instrumental in directing the behavior of others, pre-verbal children develop 'pointing' eagerly without much prompting. They appear to derive satisfaction from the sheer joining of attention with others. This hints at a telling difference at a motivational level, namely a comparatively highly developed propensity in our species for creating cooperative relationships and experiencing shared intentionality (e.g. Liszkowski 2007). This has been regarded as a crucial condition at the basis of our success as a culturally cooperative species (Tomasello et al. 2005; Fitch et al. 2010).

The restriction of non-human primate gesturing to directing and requesting functions, or – in other words – its imperative and instrumental application, has also become evident in the use of gestural symbols and lexigrams by trained apes. A critical analysis by Esteban Rivas (2006) of the recorded utterances of the chimpanzees studied by Roger Fouts and a review of the literature led him to conclude that these utterances lack the typically human aspect of narrative conversation. The apes mostly signed with an acquisitive motivation.

1.7 Combinations of signals, strings versus syntax, generative grammar

The complexity of patterning, at the level of phonemes and words, in rule governed ways, yields the seemingly endless generative capacity of the human language, its most distinguishing characteristic in comparison with animal communication systems. The ape language studies have shown that these animals, just like some birds (see above), can learn symbols and use them in communication. Claims that these apes are also able to structure their messages following rules of a simple syntax (e.g. Gardner & Gardner 1994) have been criticised for methodological imperfections and overinterpretation (Terrace 1983; Sebeok & Rosenthal 1981; Rivas 2005). Well-controlled studies such as those by Savage-Rumbaugh et al. (1986; 1998), especially those with the bonobo Kanzi, have shown that these animals do, indeed, form utterances consisting of a few different signs or symbols strung together in a way which is dependent on the context. Thus Kanzi, who readily used lexigrams and manual signs, and even understood some spoken words of English, appeared to order his symbols, although in rather short utterances, following some simple rule (e.g. Savage-Rumbaugh et al. 1986, 1998).

Yet it remains questionable whether such strings derive a specific message content from a particular syntactic structure. There might be some iconic sequencing in the sense that a subject indicator tends to precede an action or an object indicator (or reversely; e.g. Schusterman & Gisiner 1988). This could undoubtedly be the beginning of an ontogenetic ritualization in which the regularities of production and understanding during social exchange would solidify into syntactical rules. So far there is no evidence that the observed regularities in sequences can be equated with the existence of syntactic rules in the production of symbol strings, neither in non-human mammals nor in birds. The same holds for the vocal and gestural utterances produced by animals living in their natural environments.

Lately there has been an increasing number of both observational studies and play-back experiments which investigate whether more complex call sequences carry different messages depending on how different call types are combined. Campbell's guenons for instance have a vocabulary of six alert calls that as a rule are uttered in series of 25 calls on average (Ouattara et al. 2009a, 2009b). By modifying the combinations and the order of calls in a combination the messages are changed. The authors discovered that specific combinations are informative about specific events, such as social events (group movement initiatives, group encounters) and the nature of threats (falling tree or predator) and type of predator (leopard, eagle). Similar conclusions have been drawn from some studies on other species (putty-nosed monkeys: Arnold & Zuberbühler 2006, 2008; gibbons: Clarke et al. 2006). Again this could be the starting point for developing syntactical rules.

Also gestures can be given in series. However, in an investigation of gorilla gestures, Genty & Byrne (2010) found no indications that combinations or extensions of gestures between conspecifics change the message or its effectiveness.

Facial displays can be combined with gestures as well. The meaning, which a social partner attributes to a hitting intention movement, is very different depending on whether it is made in combination with a 'relaxed open-mouth face' (the primate play face; van Hooff 1972) or not. This is what Bateson (1972) called 'metacommunication'; the play face is a message about the message of the gesture.

2. To conclude

The existence of a comparatively flexible repertoire of gestures in the great apes has led to the hypothesis that human language had its origin in the emancipation of gestural referential symbols. Arguments in favour are the greater flexibility of gestures seen in the natural communication of apes and the fact that they seem to be under greater voluntary control, whereas facial and vocal displays appear to be closely linked to emotional states. For a number of primate species it has been shown that they can be taught a symbolic communication system based on gestures and also on visual symbols or lexigrams. Whereas some bird species can learn to use the vocal channel for this function, apes have failed. All this has favoured a gestural hypothesis for language evolution.

However, in as far as the vocal and gestural theories of language evolution have been presented as alternative explanations, the contrast surely is not as great as was thought.

Although facial-vocal signals are often of an emotional-reflexive nature, these can also be given, like gestural signals, in a purposively regulated manner, responding to audience effects and adjusting to the attentional state of the recipient.

The evidence for a referential use of symbols under natural conditions is even greater for calls than for gestures. An interesting question is how the phonetic quality of the different grunts is related to the message. Whereas the referential calls of guenons are both fully arbitrary and species-typical, it is unknown whether the grunts of chimpanzees have somehow an iconic character. Anyway, here seems to lie a point of origin for the development of discrete arbitrary symbols.

There is some regularity in the ways sign units can be combined in complexes. This has been shown in "language" trained animals. The regularity may be an iconic representation of the depicted process, but there is no convincing evidence that this has crystallized into an internally formalised syntax. But again, it might be the syntax precursor. Naturally occurring variation in the combination of symbol elements that is associated with different messages has so far only been noted for vocalizations.

All in all, this means that a choice between a vocal and a gestural theory of language evolution is difficult to make. Perhaps we should not even try, because the two are not mutually exclusive (e.g. Arbib et al. 2008).

Literature

Andrew, Richard J. 1963. "The origin and evolution of the calls and facial expressions of the primates." *Behaviour* 20: 1–109.

Arbib, Michael A., Liebal, Katja and Pika, Simone. 2008. "Primate vocalisation, gesture and the evolution of language." *Current Anthropology* 49: 1052–1075.

Arnold, Kate and Zuberbühler, Klaus. 2006. "The alarm-calling system of adult male putty-nosed monkeys, *Cercopithecus nictitans martini*." *Animal Behaviour* 72: 643–653.

Arnold, Kate, Pohlner, Yvonne and Zuberbühler, Klaus. 2008. "A forest monkey's alarm call series to predator models." *Behavioral Ecology and Sociobiology* 62: 549–559.

Bateson, Gregory. 1972. *Steps to an Ecology of Mind: Collected Essays in Anthropology, Psychiatry, Evolution, and Epistemology*. Chicago: University of Chicago Press.

Bout, Nicolas and Thierry, Bernard. 2005. "Peaceful meaning for the silent bared-teeth displays of mandrills." *International Journal of Primatology* 26: 1215–1228.

Bradbury, Jack W. 2003. "Vocal communication in wild parrots." In *Animal Social Complexity*, Frans B. M. de Waal and Peter L. Tyack (eds), 293–321. Cambridge MA: Harvard University Press.

Cartmill, Erica A. and Byrne, Richard W. 2007. "Orangutans modify their gestural signaling according to their audience's comprehension." *Current Biology* 17: 1345–1348.

Cartmill Erica A. and Byrne, Richard W. 2010. "Semantics of orangutan gesture: determining structure and meaning through form and use." *Animal Cognition* 13(6): 793–804.

Chomsky, Noam. 1982. *Language and the study of mind*. Cambridge MA: MIT Press.

Clarke Esther, Reichard, Ulrich H. and Zuberbühler, Klaus. 2006. "The syntax and meaning of wild gibbon songs." *PloS ONE* 1: 1–10.

Clay, Zanna and Zuberbühler Klaus. 2009. "Food associated calling sequences in bonobos *Pan paniscus*." *Animal Behaviour* 77: 1387–1396.

Conradt, Larissa and Roper, Timothy J. 2005. "Consensus decision making in animals." *Trends in Ecology and Evolution* 20: 449–456.

Corballis, Michael C. 2009. "Language as gesture." *Human Movement Science* 28: 556–565.

Darwin, Charles, 1872. *The expression of the emotions in animals and man*. London: John Murray.

Deacon, Terrence W. 1998. *The symbolic species: The co-evolution of language and the brain*. New York: Norton.

Eibl-Eibesfeldt, Irenäeus and Sütterlin, Christa. 1992. *Im Banne der Angst. Zur Natur- und Kunstgeschichte menschlicher Abwehrsymbolik*. München: Piper.

Ekman, Paul and Friesen, Wallace V. 1975. *Unmasking the face*. Englewood Cliffs, NJ: Prentice-Hall.

Evans, Christopher S. and Evans, Linda. 2007. "Representational signalling in birds." *Biology Letters* 22: 8–11.

Fitch, Tecumseh W., Huber, Ludwig and Bugnyar, Thomas. 2010. "Social cognition and the evolution of language: Constructing cognitive phylogenies." *Neuron* 65: 795–814.

Fouts, Roger S. 1975. "Capacities for language in great apes." In *Socioecology and Psychology of Primates*, Russell H. Tuttle (ed.), 371–390. The Hague: Mouton.

Frijda, Nico H. 1987. *The Emotions*. Cambridge: Cambridge University Press.

Gardner, R. Allen and Gardner, Beatrice T. 1969. "Teaching sign language to a chimpanzee." *Science* 165: 664–672.

Gardner, Beatrice T. and Gardner, R. Allen. 1994. "Development of phrases in the utterances of children and cross-fostered chimpanzees. In *The Ethological Roots of Culture*, R. Allen Gardner, Beatrice T. Gardner, Brunetto Chiarelli and Frans X. Plooij (eds), 223–255. Dordrecht: Kluwer Academic.

Genty, Emilie, Breuer, Thomas, Hobaiter, Catherine and Byrne, Richard W. 2009. "Gestural communication of the gorilla (*Gorilla gorilla*): Repertoire, intentionality and possible origins." *Animal Cognition* 12: 527–546.

Genty, Emilie and Byrne, Richard W. 2010. "Why do gorillas make sequences of gestures?" *Animal Cognition* 13: 287–301.

Hallberg, Karin L., Nelson, Douglas A. and Boysen, Sarah T. 2003. "Representational vocal signaling in the chimpanzee." In *Animal Social Complexity*, Frans B. M. de Waal and Pheter L. Tyack (eds), 317–321. Cambridge MA: Harvard University Press.

Hayes, Keith J. and Hayes, Catherine. 1952. "Imitation in a home-raised chimpanzee." *Journal of Comparative and Physiological Psychology* 45: 450–459.

Hewes, Gordon W. 1973. "Primate communication and the gestural origins of language." *Current Anthropology* 14: 9–10.

van Hooff, Jan A. R. A. M. 1962. "Facial expressions in higher primates." *Symposia of the. Zoological Society of London* 8: 97–125.

van Hooff, Jan A. R. A. M. 1967. "The facial displays of the catarrhine monkeys and apes." In *Primate Ethology*, Desmond Morris (ed.), 7–68. London: Weidenfeld & Nicolson.

van Hooff, Jan A. R. A. M. 1972. "A comparative approach to the phylogeny of laughter and smiling." In *Non-verbal Communication*, Robert A. Hinde (ed.), 209–241. London: Cambridge University Press.

van Hooff, Jan A. R. A. M. 1973. "A structural analysis of the social behaviour in a semi-captive group of chimpanzees." In *Expressive Movement and Nonverbal Communication*, Mario von Cranach & Ian Vine (eds), 209–241. London: Academic Press.

van Hooff, Jan A. R. A. M. 1976. "The comparison of facial expressions in man and higher primates." In *Methods of Inference from Animal to Human Behaviour*, Mario von Cranach (ed.), 165–196. The Hague/Chicago: Mouton-Aldine.

von Frisch, Karl. 1965. *Tanzsprache und Orientierung der Bienen*. Berlin: Springer.

Kuhl, Patricia K. 2000. "A new view of language acquisition." *Proceedings of the National Academy of Sciences of the United States of America* 97: 11850–11857.

Kummer, Hans. 1996. *Weiße Affen am Roten Meer, das soziale Leben der Wüstenpaviane*. München: Piper.

Leavens, David A. 2012. "Pointing: Contexts and instrumentality." In *Developments in Primate Gesture Research*, Simone Pika and Katja Liebal (eds), 179–195. Amsterdam/Philadelphia: John Benjamins Publishing Company.

Leavens, David A. and Hopkins, William D. 2007. "Multimodal concomitants of manual gesture by chimpanzees (*Pan troglodytes*): Influence of food size and distance." In *Gestural communication in human and nonhuman primates*, Katja Liebal, Cornelia Müller and Simone Pika (eds), 69–82. Amsterdam/Philadelphia: John Benjamins Publishing Company.

Liebal, Katja, Call, Josep and Tomasello, Michael. 2004. "Chimpanzee gesture sequences." *Primates* 64: 377–396.

Liebal, Kristin, Behne, Tanya, Carpenter, Malinda and Tomasello, Michael. 2009. "Infants use shared experience to interpret pointing gestures." *Developmental Science* 12: 264–71.

Lieberman, Philip, 2006. *Toward an evolutionary biology of language*. Cambridge: Harvard University Press.

Lieberman, Philip, Klatt, Dennis H. and Wilson, William H. 1969. "Vocal tract limitations on the vowel repertoires of rhesus monkey and other nonhuman primates." *Science* 164: 1185–87.

Liszkowski, Ulf. 2007. "Human twelve-month-olds point cooperatively to share interest with and helpfully provide information for a communicative partner." In *Gestural communication in human and nonhuman primates,* Katja Liebal, Cornelia Müller and Simone Pika (eds), 121–138. Amsterdam/Philadelphia: John Benjamins Publishing Company.

Lorenz, Konrad and Tinbergen, Niko. 1938. "Taxis und Instinkthandlung in der Eirollbewegung der Graugans." *Zeitschrift für Tierpsychologie* 2, 61–98.

Lorenz, Konrad. 1943. "Die angeborenen Formen möglicher Erfahrung." *Zeitschrift für Tierpsychologie* 5: 235–409.

Marentette, Paula and Nicoladis, Elena. 2012. "Does ontogenetic ritualization explain early communicative gestures in human infants?" In *Developments in Primate Gesture Research*, Simone Pika and Katja Liebal (eds), 33–53. Amsterdam/Philadelphia: John Benjamins Publishing Company.

Matsuzawa, Tetsuro. 1985. "Color naming and classification in a chimpanzee (*Pan troglodytes*)." *Journal of Human Evol*ution 14: 283–291.

Matsuzawa, Tetsuro. 2003. "The Ai project: Historical and ecological contexts." *Animal Cognition* 6: 199–211.

McGrew, William C., Marchant, Linda F., Scott, S. E. and Tutin, Caroline E. G. 2001. "Intergroup differences in a social custom of wild chimpanzees: The grooming hand-clasp of the Mahale Mountains." *Current Anthropology* 42: 148–153.

Menzel, Charles. 2005. "Expressive behaviors shown by a chimpanzee during her recall and reporting of hidden objects." *International Conference and Workshop "The Chimpanzee Facial Action Coding System,"* Portsmouth UK, March 2005.

Ortony, Andrew and Turner, Terence J. 1990. "What's basic about basic emotions?" *Psychological Review* 97: 315–331.

Ouattara, Karim, Lemasson Alban and Zuberbühler Klaus. 2009a. "Campbell's monkeys concatenate vocalizations into context-specific call sequences." *Proceedings of the National Academy of Sciences* 106: 22026–22031.

Ouattara, Karim, Lemasson Alban and Zuberbühler Klaus. 2009b. "Campbell's monkeys use affixation to alter call meaning." *PloS ONE* 4(11): e7808.

Pepperberg, Irene. 2000. *The Alex Studies: Cognitive and communicative abilities of grey parrots.* Cambridge MA: Harvard University Press.

Perlman, Marcus, Tanner, Joanne E. and King, Barbara J. 2012. "A mother gorilla's variable use of touch to guide her infant: Insights into iconicity and the relationship between gesture and action". In *Developments in Primate Gesture Research*, Simone Pika and Katja Liebal (eds), 55–71. Amsterdam/Philadelphia: John Benjamins Publishing Company.

Petitto, Laura A., Holowka, Siobhan, Sergio, Lauren, Levy, Barbara and Ostry, David. 2004. "Baby hands that move to the rhythm of language: Hearing babies acquiring sign languages babble silently on the hands." *Cognition* 9: 43–73.

Pika, Simone and Liebal, Katja (eds). 2012. "Introduction: Developments in primate gesture research". In *Developments in Primate Gesture Research*, Simone Pika and Katja Liebal (eds), 1–11. Amsterdam/Philadelphia: John Benjamins Publishing Company.

Pika, Simone, Liebal, Katja and Tomasello, Michael. 2003. "Gestural communication in young gorillas: Gestural repertoire and use." *American Journal of Primatology* 60: 95–111.

Pika, Simone, Liebal, Katja, Call, Josep and Tomasello, Michael. 2007. "The gestural communication of apes." In *Gestural communication in human and nonhuman primates.* Katja Liebal, Cornelia Müller and Simone Pika (eds), 37–51. Amsterdam/Philadelphia: John Benjamins Publishing Company.

Pika, Simone and Mitani, John. 2006. "Referential gestural communication in wild chimpanzees (*Pan troglodytes*)." *Current Biology* 16: 191–192.

Plutchik, Robert. 1980. "A general psychoevolutionary theory of emotion." In *Emotion: Theory, research, and experience: Vol. 1. Theories of Emotion,* Robert Plutchik and Henry Kellerman (eds), 3–33. New York: Academic Press.

Pollick Amy S. and de Waal, Frans B. M. 2007. "Ape gestures and language evolution." *Proceedings of the National Academy of Sciences of the United States of America* 104 (19): 8184–8189.

Premack, David. 1971. "On the assessment of language competence in the chimpanzee." In *Behavior of Nonhuman Primates*, Allan M. Schrier and Fred Stolnitz (eds), Vol. IV. 185–228. New York: Academic Press.

Preuschoft, Signe. 1995. *'Laughter' and 'Smiling' in Macaques-An Evolutionary Perspective.* Utrecht: Universiteit Utrecht.

Preuschoft, Signe and van Hooff, Jan A. R. A. M. 1997. "The social function of 'smile' and 'laughter': Variations across primate species and societies." In *Nonverbal communication: Where nature meets culture*, Ullica Segerstråle and Peter Molnár (eds), 171–191. Mahwah NJ: Lawrence Erlbaum.

Rivas, Esteban. 2005. "Recent use of signs by chimpanzees (*Pan troglodytes*) in interactions with humans." *Journal of Comparative Psychology* 119: 404–417.

Rizzolatti, Giacomo and Arbib, Michael A. 1998. "Language within our grasp." *Trends in Cognitive Science* 21: 188–194.

Rizzolatti, Giacomo and Sinigaglia, Corrado. 2008. *Mirrors in the brain.* Oxford: Oxford University Press.

Rumbaugh, Duane M. 1977. *Language learning by a chimpanzee: The LANA project.* New York: Academic Press

Rüppelll, Georg. 1969. "Eine Lüge als gerichtete Mitteilung beim Eisfuchs." *Zeitschrift für Tierpsychologie* 26: 371–374.

Savage-Rumbaugh, Sue, McDonald, Kelly, Sevcik, Rose A. and Hopkins, William D. 1986. "Spontaneous symbol acquisition and communicative use by pygmy chimpanzees (*Pan paniscus*)." *Journal of Experimental Psychology* 115: 211–235.

Savage-Rumbaugh, E. Sue, Shanker, Stuart G. and Taylor, Talbot J. 1998. *Apes, language and the human mind.* Oxford: Oxford University Press.

Schusterman, Ronald J. and Gisiner, Robert. 1988. "Artificial language comprehension in dolphins and sea lions: The essential cognitive skills." *Psychological Record* 38: 311–348.

Sebeok, Thomas A. and Rosenthal, Robert. 1981. *The clever Hans phenomenon: Communication with horses, whales, apes, and people.* New York: The New York Academy of Sciences.

Seyfarth, Robert M., Cheney, Dorothy L. and Marler, Peter. 1980. "Monkey responses to three different alarm calls: Evidence for predator classification and semantic communication." *Science* 210: 801–803.

Short, Lesterand and Horne, Jennifer F. M. 2002. *Toucans, Barbets and Honeyguides*. Oxford: Oxford University Press.

Slocombe, Katie and Zuberbühler, Klaus. 2005. "Functionally referential communication in a chimpanzee." *Current Biology* 15: 1779–1784.

Slocombe, Katie and Zuberbühler, Klaus. 2007. "Chimpanzees modify recruitment screams as a function of audience composition." *Proceedings of the National Academy of Sciences* 104: 17228–17233.

Slocombe, Katie and Zuberbühler, Klaus. 2010. "Vocal communication in chimpanzees." In *The Mind of the Chimpanzee: Ecological and empirical perspectives*, Elisabeth V. Lonsdorf, Steven Ross and Tetsuro Matsuzawa (eds), 192–207. Chicago: University of Chicago Press.

Stolba, Alex. 1979. *Entscheidungsfindung in Verbänden von Papio hamadryas*. Dissertation: University of Zürich, Switzerland.

Terrace, Herbert S. 1983. "Apes who talk 'language' or projection of language by their teachers?" In *Language in primates; perspectives and implications*, Judith de Luce and Hugh Wilder (eds), 19–42. New York: Springer.

Tinbergen, Niko. 1954. "The origin and evolution of courtship and threat displays." In *Evolution as a process*, Alister C. Hardy, Julian S. Huxley and Edmund B. Ford (eds), 233–250. London: Allen & Unwin.

Tomasello, Michael. 1996. "Do apes ape?" In *Social learning in animals: The roots of culture*, Cecilia M. Heyes and Bennett G. Galef (eds), 319–346. San Diego: Academic Press.

Tomasello, Michael. 2003. *Constructing a Language: A usage-based theory of language acquisition*. Cambridge MA: Harvard University Press.

Tomasello, Michael, Carpenter, Malinda and Liszkowski, Ulf. 2007. "A new look at infant pointing." *Child Development* 78: 705–722.

Tomasello, Michael and Zuberbühler, Klaus. 2002. "Primate vocal and gestural communication." In *The Cognitive Animal*, Mark Bekoff, Colin Allen and Gordon M. Burghardt (eds), 293–299. Cambridge MA: MIT Press.

Tomasello, Michael, Carpenter, Malinda, Call, Josep, Behne, Tanya and Moll, Henrike. 2005. "Understanding and sharing intentions: The origins of cultural cognition." *Behavioral and Brain Sciences* 28: 675–691.

de Waal, Frans B. M., van Hooff, Jan A. R. A. M and Netto, Willem J. 1976. "An ethological analysis of types of agonistic interaction in a captive group of Java-monkeys (*Macaca fascicularis*)." *Primates* 17: 257–290.

de Waal, Frans B. M. 1982. *Chimpanzee politics. Power and sex among apes*. London, Jonathan Cape/New York, Harper and Row.

Zuberbühler, Klaus, Ouattara, Karim, Bitty, Anderson, Lemasson, Alban and Noë, Ronald. 2009. "The primate roots of human language: Primate vocal behaviour and cognition in the wild." In *Becoming eloquent: Advances in the emergence of language, human cognition, and modern cultures*, Francesco d'Errico and Jean-Marie Hombert (eds), 235–264. Amsterdam/Philadelphia: John Benjamins Publishing Company.

Does ontogenetic ritualization explain early communicative gestures in human infants?

Paula Marentette[1] & Elena Nicoladis[2]
[1]University of Alberta–Augustana, Camrose, Alberta, Canada / [2]University of Alberta, Edmonton, Canada

This chapter evaluates predictions that the origins of at least some gestures of human infants are shaped through ontogenetic ritualization (Clark 1978; Lock 1978; Tomasello 2008). To do so, we examine the development of five gesture types in four children between the ages of 6.5 and 11.5 months. This analysis supports a comparison between those gestures thought to be ritualized (*pick-me-up, grasp-index*), with those that are socially learned (*wave, clap, clasp-hands*). Although some limited evidence for ritualization is found, developmental similarity amongst the five gesture types is strong. Through comparisons of gesture ontogeny between human and non-human primates, we explore the possible reasons for this result.

1. Introduction

A key piece of information that may increase understanding of primate communication and, in particular, the role of gesture, lies in its ontogeny. Gestures are intentional and communicative movements of the hands. Clark (1978) argued that children's earliest gestures emerge from having participated in actions; for example, the *give-me* gesture could come from a child's reaching toward an object and a parent's interpretation of that reach as a request for that object. Similar claims were made by Lock (1978) regarding the *pick-me-up* gesture, a request made by extending both arms toward a person with the intention of being moved to another position. The purpose of this chapter is to examine the development of children's early manual activities that may become communicative gestures, particularly in the 6–10 month window where there is little extant longitudinal data. In particular, this chapter will attempt to evaluate predictions that the origins of at least some gestures of human infants are shaped from functional, but not yet communicative, movements. This process is an individual learning process called ontogenetic ritualization (Clark 1978; Lock 1978; Tomasello 2008), which may also play a crucial role in the acquisition of gestural signaling in great apes (Tomasello 1996; Pika, Liebal, Call & Tomasello 2005). Understanding the origins of early infant action/gesture, and the similarities and differences from the origins of early gestures of non-human primates will advance our knowledge of primate communication and the origins of human language.

To our knowledge, the assumption that ontogenetic ritualization explains the origins of at least some children's gestures has not been tested. In this chapter we will explore two explanations for the origins of gesture in human infants: ontogenetic ritualization and social learning (see Carpenter & Call 2007). Looking first at explanations for *pick-me-up*, a gesture that is claimed to be a result of ontogenetic ritualization and then at conventional gestures including *clap* and *wave* that are thought to be the result of social learning, we will examine data from case studies of four children. The chapter begins with a brief review of three approaches to gestural origins in non-human primates followed by a review of claims of continuity between action and gesture in human infants. We will conclude with a discussion of the challenges of working with longitudinal data in humans as well as the implications of these results for studies of the origins of human communication.

1.1 Gestural origins in non-human primates

Gestures in non-human primates are thought to have three possible sources: (a) ontogenetic ritualization (Call & Tomasello 2007), (b) social learning, such as imitation (Pika, Liebal & Tomasello 2003; 2005) or (c) species-typical behavior (Byrne & Tanner 2006; Genty Breuer Hobaiter & Byrne 2009). Tomasello (1996) identified the key mechanism by which chimpanzees and other non-human primates learn a wide variety of gestures as ontogenetic ritualization (see also Pika et al. 2003; 2005; Tanner & Byrne 1996). An action that an individual ape produces, typically as part of a movement sequence, over time comes to elicit the whole sequence of actions. Importantly, Tomasello and colleagues (1997) claim these gestures are not acquired through imitation, a more complex form of social learning. Ontogenetic ritualization is a simpler learning process that primarily requires an ability to anticipate what a conspecific will do next.

The marks of a gesture acquired through ontogenetic ritualization include: (a) a form that is derived from an associated mechanically effective action, (b) role and dyad specificity, and (c) a high degree of individual variation (Tomasello et al. 1997). The form of a ritualized gesture should be identifiable as one element of a more complex action but not one that could achieve the desired goal on its own. According to Genty et al. (2009), one implication of ontogenetic ritualization is that these gestures ought to be distinct to each individual's role in the dyad, rather than generalized across individuals and groups. A developmental course is also implied (Tomasello 1996). Initially, one participant is the initiator and the other participant the receiver. With repeated performances of the action-sequences, the initiator performs a specific part of the action in order to elicit a response from a particular receiver (see Clark 1978 who outlines the same process for human infants). A corollary of the nature of ontogenetic ritualization and the degree of dyadic specificity is a high degree of individual variation in gesture form (Tomasello et al. 1997).

According to Tomasello et al. (1997), ontogenetic ritualization is the best explanation of the majority of non-human primate gesture because of the wide range of individual variation observed in gesture production and the very limited examples of group specificity. Call & Tomasello (2007) demonstrate that across five species of primate, the typical individual gesture repertoire shows a high degree of selectivity from the repertoire of the species as a whole. If a form of social learning were to be the primary influence on gesture, then one would expect to see group-specific gestural repertoires and would not expect to see: (a) idiosyncratic gestures, (b) unidirectional gestures produced, for example, only by juveniles toward adult conspecifics, or (c) species-typical gestures produced by isolated individuals (Call & Tomasello 2007). In addition, ontogenetic ritualization makes a more conservative claim about the cognitive skills of the individual. Ontogenetic ritualization requires only that the juvenile ape can anticipate that a certain movement in a certain context will lead to fulfillment of a pre-existing goal. Imitation requires that one partner recognizes the intention of the other partner for the communication to succeed (Tomasello 2008).

While acknowledging the importance of ontogenetic ritualization, Arbib Liebal & Pika (2008) argue that there is a role for social learning in the form of simple imitation in the gestural repertoire of non-human primates. In particular, they argue that there is some evidence for group-specific gestures. In contrast, Byrne & Tanner (2006) provide evidence that novel gestural forms are not learned by imitation in gorillas; instead, similar but rare forms in their gestural repertoire are facilitated by the modeling of "novel" gestural forms.

In addition, Genty et al. (2009) find little evidence to support social learning processes such as imitation in their analysis of gorilla gestural repertoire. They also argue against ontogenetic ritualization as a primary source of gestures among gorillas. Instead, Genty and colleagues claim that gorillas have a large repertoire of species-typical actions that are recruited in flexible and intentional ways for communication. This is in contrast with Call & Tomasello (2007), who assume that species-typical gestural forms are inflexible in their use across contexts since they are not learned. Genty et al. reject ontogenetic ritualization as an explanation since they were not able to find evidence that would provide unique support: there were no examples of dyad-specific gesture use, and species-typical forms with no obvious basis in an action sequence were used flexibly and with intention across contexts.

In sum, ontogenetic ritualization is generally (though not uniformly) accepted as a primary form of learning for gesture in non-human primates (e.g. Arbib et al. 2008; Call & Tomasello 2007). The alternatives are disputed. Social learning processes such as imitation are, at best, agreed to be limited to the few gestures that occur uniquely in particular groups (e.g. Arbib et al. 2008). The role of species-typical forms is also contested (Call & Tomasello 2007; Genty et al. 2009); the dispute is primarily based on whether or not genetically fixed behaviors can be flexible in their use.

We now turn to the exploration of gestural origins in human primates. The literature in this area has suggested a role for ritualization, but assumes that most gestures are conventional, sharing a fixed form with a specific meaning within a group that is learned by the child through social interaction.

1.2 Gestural origins in human primates

Do children ritualize action into gesture? If so, how does this process compare with the development of socially meaningful gestures that children produce? Tomasello (2008), citing Lock (1978), claims that human infants learn at least some of their gestures, notably the *pick-me-up* gesture, through ontogenetic ritualization. Clark (1978) argued that early communication in human infants resulted from direct action altered by social interaction into the typical gestures and forms used by adults (see also Blake 2000; Caselli 1990). According to Clark, the transformation of action into gesture was marked by the advent of the explicit distinction made by the child between the action to achieve the goal and the gesture to indicate *intention* to achieve it. The mother supported this development by responding to the action *as if* it were communicative (see also Erikson 1966). Zinober & Martlew (1985) also make the distinction between acts, which are performed with a focus on an outcome goal or event, and gestures, whose communicative intention leads to a focus on the dyadic partner. They reported that mothers of infants respond differently to gestures such as *pick-me-up*, which "may well be shaped by the responsiveness of the mother", and enactive gestures that were explicitly molded and modeled (such as those that occur in rhyming games with conventional actions, Zinober & Martlew 1985, p. 304). Further, Tomasello (2008) argued that conventional language arises out of the mastery of gesture and the cultural learning skills that result from participating in social and cooperative contexts. It is the potential link between early gesture and language outcomes (both ontogenetic and phylogenetic) that lends importance to understanding the origins of early human gestures. Do intentional gestures originate in movement sequences that are functional but not communicative? We know of no studies presenting data that evaluate this claim.

Perhaps the clearest candidates for such continuity are empty-handed manual movements that could have originated from being a participant in an action (such as *pick-me-up*, or *give-me*). However, similar arguments have been made about *pointing* (see Blake 2000). Previous research looking for continuity between early actions and later gestures has led to various claims. Part of the reason for the variety of claims could be the way researchers have structured their observations. According to the Communicative Development Inventory (CDI), a parental report language development measure, more than 80% of children produce the *pick-me-up* gesture "often" by 8 months of age (Dale and Fenson 1996). However, most researchers begin their studies with 9-month-olds (e.g. Blake 2000; Masur 1983; Messinger & Fogel 1998; Zinober & Martlew 1985), at which time some early gestures may already be acquired.

1.3 This study

The purpose of this study was to assess the extent to which ontogenetic ritualization explains the development of early gestures, in particular the transfer gesture *pick-me-up.* This development will be compared to that of early conventional gestures *wave* and *clap.* Over the course of the study two other gestures, both infrequent and limited to a subset of children, were noted and their patterns are also reported. These include *grasp-index*, a potentially ritualized gesture that facilitates the infant's movement to a standing position, and *clasp-hands*, a culturally-specific conventional gesture. (All gestures are described below in **Methods**). This descriptive study followed four infants longitudinally from 6.5 months to 11.5 months, an age that spans the period generally recognized as the onset of intentional communicative behavior (e.g. Tomasello 2008) and begins several months earlier than most other data sets. The children came from two different cultural backgrounds: two were French Canadian and two were Chinese Canadian.

Gestures emerging from ontogenetic ritualization should show evidence of gradual emergence from functional action sequence to communicative request, be unidirectional (role-specific) and dyad specific, and have high individual variation in form. They may also occur in a limited range of contexts (Call & Tomasello 1997). Following Lock (1978) and Tomasello (2008), we predict that these features will be found in the development of *pick-me-up* and *grasp-index*.

In contrast, we expect conventional gestures, such as *clap*, *wave*, and *clasp-hands*, to show evidence of social learning processes, in particular imitation. In this case we expect to observe: stable form, rapid emergence, use of the same gesture form by both members of the dyad, and use with multiple partners. Because the form of conventional gestures varies with culture (McNeill 1998), we expected that the different cultural backgrounds of the children would lead to the production of different conventional gestures. This expectation was borne out given our observation of the gesture *clasp-hands*.

2. Methods

2.1 Participants

Four children and their families participated in this study. All of the children lived with their mothers and fathers. Two of the children had siblings (KV about 9 years older; ZN about 5 years older). All children were growing up in Edmonton, Alberta, Canada, a predominantly English-speaking part of Canada. Two of the children (AD and KV) regularly heard Mandarin Chinese as well as English. Both children's parents were native speakers of Mandarin Chinese and spoke English fluently. Two of the children (JS and ZN) heard French as well as English on a regular basis. In both cases, the children's mothers spoke only French with the children and their fathers were native speakers of English. The fathers spoke French fluently.

2.2 Procedure

The children were filmed over two days approximately every two weeks, from the time they were 6.5 months to 11.5 months, unless the family was unavailable. The families were asked to film representative samples of the children's everyday life. The situations of each home varied considerably, both by family and by filming session, including situations such as mealtime, bathing, and play. The duration of each session was 2 hours. Total film length ranged from 15 to 22 hours across children. The ages of the children and the minutes coded at each age for this study are shown in Table 1.

2.3 Coding gestures

We coded the frequency of occurrence of the following gesture types: (a) *pick-me-up*, a behavior involving the extension of the arms toward an adult, typically with open hands, with the goal of being picked up under the arms and moved from the current position; (b) *grasp-index*, a behavior involving an adult extending his or her index fingers in front of the child, for the child to grasp and use as leverage

Table 1. Age, minutes coded, frequency and distribution of gestures by gesture type

	Age (in months)														
	Child	6	6.5	7	7.5	8	8.5	8.75	9	9.5	10	10.5	11	11.5	Total
Minutes coded	C-AD	–	60	60	60	60	60	60	60	60	45	45	45	–	615
	C-KV	–	60	60	60	60	–	–	60	60	60	–	60	60	540
	F-JS	–	–	60	60	60	60	60	42	–	60	–	–	60	462
	F-ZN	60	60	60	60	–	–	–	–	60	60	–	60	60	480
pick-me-up	C-AD			2	1	10	14		5	1	3	2	3		41
	C-KV					2				4	4		2	3	15
	F-JS						1	2	4		1			2	10
	F-ZN	1				1				5	1		5		13
grasp-index	C-AD			1	1	6	1		2						11
	F-ZN												1		1
clap	C-AD									10			2		12
	C-KV				2					2	2		9	4	19
	F-JS										19				19
	F-ZN			2	25	2				1			1		31

(*Continued*)

Table 1. Age, minutes coded, frequency and distribution of gestures by gesture type (Continued)

		Age (in months)													
	Child	6	6.5	7	7.5	8	8.5	8.75	9	9.5	10	10.5	11	11.5	Total
clasp-hands	C-AD								12	7	1		5		25
	C-KV			4	1	3			1	1			5	2	17
	F-JS						1				2				3
	F-ZN		1	1	2						2			1	7
wave	C-AD		2				5			6					13
	C-KV				1						9		10	1	21
	F-JS				1									6	7
	F-ZN				3						2		20		25

with the goal of moving from sitting to standing, this sequence was sometimes initiated by the child by holding his slightly open hands by his shoulders; (c) *wave* a conventional gesture whose target form involved the waving of fingers or wrist toward a person who was leaving; (d) *clapping*, a conventional gesture whose target form involves making a noise by (repeatedly) bringing the palms of the hands together in the saggital plane, typically used to express pleasure about an event or achievement; and (e) a culturally specific form of the conventional gesture *clap*, *clasp-hands*, which was encouraged in the Chinese-speaking families. The gesture is a form of respect used to acknowledge elders, the target form involves one hand clasping the other with a repeated small movement in the saggital axis. Pictures of all five gestures can be seen in Table 1.

All forms examined have as their target empty-handed forms meant to be used in dyadic rather than object-oriented situations. We coded each instance for its form, the communicative intention of the child, the nature of dyadic interaction, and the context of the child's production. In the interests of exploring the origins of each gesture type, we included all action-based forms, regardless of communicative or intentional production, that matched *either* the form or the function of the gestures described above. This overly inclusive approach to coding was intended to capture early but poorly produced instances of a form that might reveal a developmental trend.

2.4 Form

To determine form, we coded the form of the hand (open hand, palm down open, palm up open, index extension, partially closed fingers, fist, clasp-hands), the movement (clap, flap, roll fingers, open/close fingers), and whether or not the arm(s) was/were extended (hand further than an elbow's length from the body). Repeated gestures were marked as a string; only the first gesture in a string of a repeated type less than 5 seconds apart was therefore counted.

2.5 Communicative intention

To determine whether productions were intentional or not, the following aspects of each instance were coded: direction of eye gaze (object, person, unknown/off-camera), whether or not the child produced an accompanying vocalization, whether or not another individual was present, and whether or not the child was in view of that individual. These last two help to establish whether the child was producing gestures when there was no possibility of communication. It was possible, and for some children frequent, for another person to be in the room, but not attending to the child (e.g. back to child while preparing food).

2.6 Dyadic interaction

To determine the dyadic nature of the gestures, we coded whether the child initiated the form (spontaneous, imitated, elicited: a form was considered imitated if an adult produced the form immediately prior to the child's production, and elicited if the child was verbally instructed to produce the form). We also coded whether the child was persistent in achieving the goal of its production (as indicated by repeating the gesture or other bodily markers such as leaning or crying). Persistence marked intentionality but also an expectation of dyadic response. In addition, we marked whether or not there was a dyadic response (talking, achieved, refused, none) as an indication of whether the partner acknowledged the gesture as communicative.

2.7 Context

To establish flexibility of gesture use, we coded the following four contexts that occurred in these observations: (a) playing, any instance where the child was involved with toys; (b) feeding, any instance where the child was being involved with food (often but not always in the high chair); (c) sitting/standing/lying, the child was on its own, no toys or food were present; and (d) held, in these situations the child was in the arms of an adult. No instances of the gesture types of interest occurred during book reading. One instance of a *wave* gesture occurred during bathing for JS; this was collapsed into the sitting category.

2.8 Reliability

Interrater reliability was calculated on 375 (18%) of the 2100 minutes of coding. An interrater reliability of 77.5% was achieved for agreement on events occurring. A Cohen's kappa of .866 was achieved for agreement about event type.

3. Results

A total of 290 forms were observed that fit into the five gesture types studied. The children varied in their frequency of production of gestures (AD n = 102; KV n = 72; JS n = 39; ZN n = 77). We report the results for each gesture type through an analysis of the following features: (a) frequency of production by age (see Table 1), (b) intentionality of production, (c) stability or variability of the form as well as its unidirectional (form specific to each partner) or bidirectional nature (form shared by partners), (d) context of production, and (e) dyadic interaction. These features allow us to examine the data in light of the predictions regarding the ontogeny of ritualized and conventional gestural forms.

3.1 Potentially ritualized forms

3.1.1 *Pick-me-up*

This frequently produced gesture (n = 79 across children) was used by all four children as a request to change locations. The gesture was produced with communicative intent. Only two instances of the total of 79 were produced by a child who was alone (AD n = 1; JS n = 1). Of the other 77 instances, 73 were directed at other people as assessed by eye gaze. ZN produced the earliest example of this form, occurring at six months of age. At this age, the timing was closely synchronized and it is difficult to tell who initiated the sequence. With the exception of this very early instance by ZN, the children began to produce this form during the course of the study. From its earliest use, this form was part of an action sequence whose function was apparent to both child and adult. When the children first used it spontaneously as a request to move, their directed eye gaze suggests that it was used communicatively. This first spontaneous use occurred prior to nine months for all children (AD, 7.5 months; KV, 8 months; JS, 8.75 months; ZN, 8 months). Children were often persistent in the production of this gesture (AD n = 30/41; KV n = 12/15; JS n = 4/10; ZN n = 5/13).

The form of the *pick-me-up* gesture was unidirectional, that is the form produced by the child (extended arms directed toward the adult) was not produced by the partner (whose response, if they agree to the request, was to place their hands under the child's arms). When adults elicited the *pick-me-up* form, they did not extend their arms, they placed their open hands near the child's armpits. No instances were observed of an adult producing the extended *pick-me-up* form. A particular elicitation routine around the production of the *pick-me-up* gesture was observed in AD's family. Of the five adults in AD's family, four were observed using a distinct action sequence involving clapping twice, and presenting their open hands palm up in front of him. AD responded with the *pick-me-up* gesture. At this point AD was picked up under the arms by the adult.

The form of the *pick-me-up* gesture was stable with variation limited to whether the open hand was palm down (preferred by KV n = 10/15 instances) or palms facing each other (preferred by the other children AD n = 30/41; JS n = 9/10; ZN n = 11/13). Remaining instances were the opposite form for each child (i.e. palm facing if they typically produced down) with the exception of five instances involving its production with closed or curled fingers (AD n = 3, JS n = 1, ZN n = 1). The arms were always extended, either out or up. The form appears to be quite stable and there is no evidence to suggest that the gesture changed its form developmentally over the period studied.

Since this gesture was used to request a change in location, we expected it would be produced in all contexts, possibly with the exception of the "held" context in which the child was already in a parent's arms. Only one child, AD, produced this gesture while held prior to being transferred from one adult to another. As can be seen in Table 2, the form was produced in each of the other contexts, suggesting the form was not a routine limited to a particular activity or location.

Table 2. Frequency of gesture type by child and context

	Child	Feeding	Playing	Sitting	Held
pick-me-up	C-AD	13	14	10	4
	C-KV	4	5	6	0
	F-JS	2	4	4	0
	F-ZN	7	1	5	0
grasp-index	C-AD	0	6	5	0
clap	C-AD		0	8	4
	C-KV	4	3	1	11
	F-JS	0	4	15	0
	F-ZN	4	8	16	3
clasp-hands	C-AD	7	0	12	6
	C-KV	6	4	6	1
	F-JS	0	2	1	0
	F-ZN	2	1	1	3
wave	C-AD	0	0	5	8
	C-KV	6	3	6	6
	F-JS	0	0	1	6
	F-ZN	20	0	3	2
Total		75	55	105	54

When children used this form spontaneously, it was a request to change location (see Table 3). Parents did not always grant this request, which was often part of a "negotiation" about what would happen next. For example, in the feeding context, children were often ready to get out of the highchair before adults were finished feeding them. In these cases, adults often responded by ignoring or trying to distract the child. AD was the most abundant producer of this gesture, but only seven of his 32 spontaneous productions resulted in him being picked up. All of KV's *pick-me-up* gestures were spontaneously produced, but only two of 15 were successful. Similarly eight of ten gestures for JS were spontaneous, but only one of these was successful. For ZN two of her eight spontaneous gestures resulted in a change in location. When this form was elicited by parents, children responded with the gesture, indicating their willingness to be picked up. This was not simply a mechanical interaction, as parents picked up their children in other situations without "asking" them first. These data suggest that the form is not highly rewarded.

Table 3. Frequency of gesture type by child and form of behavior

	Child	Non-dyadic	Molding	Elicited	Imitated	Spontaneous
pick-me-up	C-AD	1	0	8	0	32
	C-KV	0	0	0	0	15
	F-JS	1	0	1	0	8
	F-ZN	0	0	4	0	9
grasp-index	C-AD	0	0	7	0	4
clap	C-AD	0	0	8	1	3
	C-KV	0	8	2	2	7
	F-JS	3	0	0	10	6
	F-ZN	0	0	0	1	30
clasp-hands	C-AD	0	1	16	0	8
	C-KV	1	1	0	0	15
	F-JS	2	0	0	0	1
	F-ZN	0	0	0	0	7
wave	C-AD	0	1	11	0	1
	C-KV	0	1	5	8	7
	F-JS	0	4	1	1	1
	F-ZN	0	0	1	2	22

All children except JS used this gesture with multiple partners. During the observations, JS rarely interacted with anyone other than his mother. Given the specific eliciting routine used by four of the five adults in AD's family, it is easy to see that this form is not dyad specific. The form was used by ZN with both parents in her family and by KV with both his older sister and mother who were his most frequent companions during the observations.

3.1.2 *Grasp-index*

This form was directly related to the action sequence of being helped to standing from a sitting position. This form was only produced as a gesture by AD among the four children in this study. Once at eleven months, ZN did hold on to her mother's offered index fingers, but we observed no other instances of an action sequence or offer/request. Despite its very low frequency and its repeated production in only one family (factors that limit interpretation), this form is reported here for accuracy and to support future observations.

The action sequence was observed eleven times when AD was between the ages of seven and nine months (see Table 1). The age range in which it occurred coincided with AD's increasing physical capacity to stand up. By nine months, he began to be able to pull himself up to standing and was in less need of this support. The form does emerge gradually, at least on tape, as adults appear to shape AD's response to the offer: for example in the first observation at seven months,

AD's mother offered her index fingers and refused AD's response of *pick-me-up* (see photo in Table 1), and then put her index fingers directly in his hands. AD first spontaneously initiated this routine at eight months of age, producing it spontaneously four times in all. Two of those four instances involved his sitting with his back to his mother, so eye gaze cues were not informative. The other two instances involved eye gaze to the father. The form was clearly unidirectional with the adult extending index fingers as an offer or AD holding his open hands palm out near his shoulders as a request. This form occurred in the two contexts most likely to involve this type of movement: playing and sitting (see Table 2). The form was more frequently elicited (by three different adults) than initiated by AD (see Table 3), suggesting that it was not dyad specific.

3.2 Conventional gestures

3.2.1 *Wave*

A total of 66 instances of *wave* were observed across all four children, with frequencies and distribution by age as indicated in Table 1. The form was intentional and dyadic as evidenced by the children's eye gaze. For virtually all forms (58 of 66 instances), both hands and eyes were oriented to people, mirrors, or cameras (except for two instances by AD, two by KV, and four by ZN). In these eight instances, it was not possible to determine what or who the child was looking at off camera.

This gesture was highly variable in form within and across children: of the 60 non-molded forms, handshape ranged from open hand, either palm out or down (n = 41 instances), to curled, rolling or closed fingers (n = 19 instances); the joint involved in the movement was variably the shoulder, elbow, wrist, proximal knuckles, or none. Table 4 conveys the ways children combined these handshapes and movements.

Table 4. Handshape and joint combinations for *wave* gestures by child

Joint	No movement	Proximal knuckle	Wrist	Radio/ulnar/ elbow	Shoulder
Fingers closed					
C-AD	1	7	1		
C-KV			5		5
F-JS					
F-ZN		9			
Open hand					
C-AD			1	2	
C-KV	3		6	6	2
F-JS					3
F-ZN		2	19		

Note: Numbers do not add up to 60 as some children produced more than one movement with each gesture.

While KV produced this form in all contexts, the other children produced it while held or sitting (see Table 2). These forms were not unidirectional, as children frequently imitated forms that were modelled for their benefit (see Table 3). In six instances, mothers molded their child's hand or arm, effectively producing the form for them. This form was also verbally elicited by parents, particularly frequently for AD.

3.2.2 *Clap*

A total of 81 instances of *clapping* were observed across all four children (see Table 1). The *clapping* gesture was not communicative in every instance. For AD *clapping* typically occurred with eye gaze to other people (7 of 11 instances) whereas for KV only 2 of 11 instances occurred with eye gaze to a person. JS produced three examples of the *clap* gesture in non-dyadic situations and 7 of the remaining 16 included eye gaze to the parents. For ZN, only eleven of 31 forms were accompanied by eye gaze to other interactants, and for many instances she did not appear to be looking at a specific object or a person at all (n = 15). We observed that ZN produced this form with a wide range of conflicting emotions (while smiling, frustrated, and crying) raising the question whether there was a clear communicative intention for this gesture for this child.

The form was standard and involved open hands with contact, but not all children were able to make a noise in their early attempts. The only variation in form observed was the use of *clasp-hands* in the two French speaking children's attempts at *clapping* (JS n = 3 at 10 months; ZN n = 2 at 7.5 months). These five instances of the form *clasp-hands* were produced in sequence with a *clapping* gesture and therefore reported in the *clapping* category. Blake (2000) presented the *clapping* gesture as *clap/clasp-hands* suggesting that this variation is not unusual.

These forms were not unidirectional; as with *wave*, children imitated forms that were modelled for their benefit (see Table 3). Only KV's mother molded her child's hand or arm, doing this when she played a *clapping* game with him. This form was also verbally elicited by parents, again with particular frequency for AD. The children produced this gesture in multiple contexts as can be seen in Table 2.

3.2.3 *Clasp-hands*

This form is a culturally meaningful gesture for the Chinese families but not for the French families. This made it a particularly interesting form to examine in terms of social learning. Notably all children produced the form, but only the children in the Chinese families went on to develop the form as a communicative gesture. A total of 52 instances of *clasp-hands* were observed.

The two French children produced this form without movement and, for JS, in non-dyadic contexts. Forms reported in this category were made while sitting or playing and the adults in their environment did not comment on or respond to these forms.

This gesture unfolded quite differently in each of the Chinese families. At AD's first production at nine months, the form appeared to be quite stable. Of his 25 instances, 16 were elicited, and AD received a great deal of praise and attention when he used this form (20 instances). He did not, however, always use the form correctly. At nine months, AD did not appear to understand the particular nature of each gesture. For example, AD spontaneously and inappropriately produced the *clasp-hands* gesture while looking at his grandfather, but when he received no response he produced *pick-me-up*. Later his grandmother asked him to *wave*, but he produced *clasp-hands*. At a different time, AD's father asked him to *wave* to his mother. When AD again produced the *clasp-hands* his father said, "someone doesn't quite understand yet". AD produced this gesture with four of the five of the adults living in his house, and also with some visitors. He produced this form with eye gaze directed at the recipient (19 of 24 productions, eye gaze for the remaining five instances cannot be ascertained from the recording).

KV's family was smaller and his mother less likely to verbally elicit gestures of any type. Much like the French children and unlike AD, KV's productions of this form did not appear to be communicative; only one of the four instances produced at seven months was accompanied by eye gaze to his mother. She did not respond to the form. At 7.5 months KV's mother molded his production of the *clasp-hands* form while he was sitting on her lap. KV produced this form with eye gaze to another person three additional times (once each at nine, eleven, and 11.5 months), for a total of four of his 17 utterances. His productions never resulted in the kind of extravagant praise that AD experienced, although his mother often talked to him in response. KV's mother may not have accepted this form as *clasp-hands* since KV only produced the form with movement on one occasion (at eight months).

Note, that neither child produced an imitated version of this form (see Table 3). Due to the respectful nature of this form, it is highly unlikely that an adult would ever produce this form toward a child. Indeed, we did not observe a single interaction where the child saw anyone produce this form, not toward him, nor toward anyone else, making imitation difficult. Two examples of molding were observed, one in each Chinese family. These forms are therefore, unidirectional despite not being based on an action sequence.

4. Discussion

The data presented in this chapter aim to address two questions: (a) Are either *pick-me-up* or *grasp-index* ritualized forms? and (b) In what way does their development mimic or differ from that observed with the socially learned forms *wave*, *clap* and *clasp-hands*?

4.1 Ritualization in human gesture ontogeny

We predicted that a gesture learned through a process of ontogenetic ritualization would show adult-initiated action-sequences before the child produced the gesture. In addition, the form would be dyad-specific, meaning that a different form could emerge for each child and with each partner. Finally, the form would be unidirectional, that is, the child's form would not be produced by the adult.

Our results did not provide unambiguous evidence for ontogenetic ritualization as the origin of either the *pick-me up* or the *grasp-index* gesture. In favour of ontogenetic ritualization, both of these gestures were clearly unidirectional. Factors against ontogenetic ritualization include their use with multiple dyads across multiple contexts. For the *pick-me-up* gesture, we also observed that its form was stable across individuals (and similar to that reported elsewhere, e.g. Blake 2000). Although this form was clearly part of an action-sequence, this study provides no evidence of gradual ritualization as children moved from functional action to communication.

The *grasp-index* gesture did show development over the course of this study: AD required help to respond to the gesture at seven months (mother put index fingers on child's palm) but spontaneously produced the form at eight months. Because this gesture was limited to one family and we observed few spontaneously produced instances, we cannot draw strong conclusions from this form. It is reported here as an excellent candidate for future studies of possible ontogenetic ritualization in young human infants, particularly due to the possibility of observing a shift from functional to communicative actions.

The potentially ritualized forms were unidirectional and emerged from action sequences, but they were also stable in form and used with multiple partners across multiple contexts. We therefore found little clear evidence for ontogenetic ritualization in the present study. There are several possible reasons for our failure to find strong evidence. First, it is possible that the *pick-me-up* gesture is produced because of a motoric predisposition to raise the arms. If this were true, then like the gorillas in Genty et al. (2009) who use species-typical gestures flexibly, the infants could have learned to use this gesture flexibly. If so, then human infants are able to use action-based gestures flexibly within six to eight months after birth. A second possible reason for our failure to find supportive evidence for ontogenetic ritualization is that our data set is not fine-grained enough to observe the process of ontogenetic ritualization taking place. If so, then it is necessary for researchers to observe infants from at least four months of age and at least weekly in order for ontogenetic ritualization to be observed. A third possibility is that children display such individualized learning patterns before the onset of conventional gesture use, that a single answer to the ontogenetic origins of gestures is impossible (see also Iverson Capirci & Caselli 1994).

Fourth and finally, it is possible that ontogenetic ritualization is simply not an important source of gesture acquisition in human development. Perhaps we can't clearly see ontogenetic ritualization in even very young children because the adult of the dyad is actively intervening to model cooperative communication (Tomasello 2008). As Jakobson (1962) proposed with the early forms of children's kinship words, it may be that children's motoric predispositions are shaped by social responding. In line with this possibility are our results that the infants in this study generalized the communicative partners with whom they used the *pick-me-up* gesture. While this generalization is not thought to be possible through ontogenetic ritualization (Genty et al. 2009), it is perfectly possible through selective social responding and operant conditioning (e.g. Stokes Baer & Jackson 1974). An example from this study is AD's *pick-me-up* gesture in response to a *clap* routine by different adults: this seems like a potential candidate for having been conditioned rather than ritualized.

4.2 Social learning in human gesture ontogeny

Unlike potentially ritualized gestures, which are based in action sequences, conventional gestures are fundamentally about social situations rather than action. It is for this reason that we expect them to be learned through social means rather than ritualization (Pika et al. 2003; 2005). With key exceptions, the children's development of conventional gestures in this study appeared to be socially learned. The forms were used across both multiple dyads and contexts. The same form was produced by both dyadic partners, with the exception of *clasp-hands*, though here the restriction on its use by adults is essentially social rather than resulting from distinct roles in an action sequence. These forms were often verbally elicited from children (contributing to their status as gestures *about* communication) and were frequently imitated from communicative partners. In particular mothers occasionally molded their children's hands and arms to produce these forms, a form of assisted imitation (Zukow-Goldring & Arbib 2007) that was never provided for the gestures *pick-me-up* or *grasp-index*. The conventional forms showed an increase toward the later end of the age range, with the exception of ZN who *clapped* frequently at 7.5 months. The forms were stable across dyads and families with the exception of *wave*, which showed a great deal of variability.

According to Tomasello (2008), successful social learning requires recognition of the intention of the dyadic partner. The examples of AD's confusion around the correct context for production of *clasp-hands* are particularly revealing. AD showed his understanding that gestures are social and communicative (he produced gestures both to achieve his own goals, and in response to verbal elicitation), but he did not understand the correct occasion for the production of this fairly subtle gestural form conveying respect. This dilemma is all the more acute for AD since no one modelled the use of *clasp-hands* for him. Where he could imitate the production of the other conventional gestures *clap* and *wave*, he must discern on his own the correct cues for the production of *clasp-hands*.

4.3 Cross-species comparison

In closing, while these results are superficially like those of Genty et al. (2009) with gorillas (in that we did not find compelling evidence for ontogenetic ritualization in two gesture forms), it is difficult to come to a straightforward answer as to whether human infants share the same kind of ontogenetic origins of gesture with other primates. This apparent similarity was found, in spite of very different methodological approaches in the identification of gestures, in the two studies. It is important to note that the present study was restricted to infants in their first year of life while Genty et al. (2009) analyzed data from infant, juvenile, and adult gorillas. Tomasello (2006) argued that it is appropriate to compare pre-linguistic human infants with other primates to rule out the possible effects of language. However, ritualized gestures are thought to be a temporary communicative means in human infants, before conventional communicative means are used (Clark 1978). It is unclear what kind of developmental sequence might be hypothesized for other primates. To test the hypothesis of similar ontogenetic origins of gestures in humans and other primates, it is important to include groups from at least two species, using the same methodology to identify gestures, and over approximately the same developmental range.

In the present study, potentially ritualized forms did not follow closely expectations for ritualized forms observed in non-human primates and outlined in the literature. Although these gestures were unidirectional and based in action sequences, we did not observe evidence of a shift from functional to communicative use. Their forms were more stable and their social context broader than expected. At the same time, the gestures that we expected to be socially learned, did not wholly align with expectations either. *Clasp-hands* was unidirectional, though not action-based. The form of *wave* was not stable, varying within and across children. *Clap* may not be particularly dyadic or communicative in its use, at least for some children at these younger ages. Data from children are messy as they are sporadic and hard to interpret. This data set is not exempt from that challenge.

A key feature of ontogenetic ritualization is the ability to anticipate a desired outcome. Clearly human infants anticipate and desire a range of outcomes: some action-based (e.g. a change in position) and some social (e.g. shared observations). In the present study, potentially ritualized forms differed from socially learned forms in one key way: the form itself was unidirectional and emerged from an action sequence. It may be that early communication in human infants is based on the ability to anticipate a desired physical outcome, but given their highly social context the distinction between action and social outcomes is blurred by their dyadic partners: human adults who behave *as if* the child was already communicating thereby helping infants leap to a communicative state (Clark 1978; Erikson 1966). Given the intervention of human adults, human infants may move through the early phases of gesture ontogeny much more quickly than young non-human primates, making this developmental phase difficult to observe.

Acknowledgements

This research was supported by a grant from the Social Sciences and Humanities Research Council of Canada. We would like to acknowledge the contributions of the four families who spent many weekends chasing their children with cameras. We would like to acknowledge Hui Yin, Carrie Jensen, Lindsay Faas, Rachel Aultman, Jennifer St. Jean, and Emma Marentette for help with data collection, coding and analysis.

References

Arbib, Michael A., Liebal, Katja and Pika, Simone. 2008. "Primate vocalization, gesture, and the evolution of human language." *Current Anthropology* 49(6): 1053–1076.

Blake, Joanna. 2000. *Routes to Child Language: Evolutionary and Developmental Precursors.* Cambridge: Cambridge University Press.

Byrne, Richard W. and Tanner, Joanne E. 2006. "Gestural imitation by a gorilla: Evidence and nature of the capacity." *International Journal of Psychology and Psychological Therapy* 6 (2): 215–231.

Call, Josep and Tomasello, Michael. 2007. "Comparing the gestures of apes and monkeys." In *The Gestural Communication of Apes and Monkeys*, Josep Call and Michael Tomasello (eds), 197–219. Mahwah, NJ: Erlbaum.

Carpenter, Malinda and Call, Josep. 2002. "The chemistry of social learning." *Developmental Science* 5: 22–24.

Caselli, M. Cristina. 1990. "Communicative gestures and first words." In *From gesture to language in hearing and deaf children,* Virginia Volterra and Carol J. Erting (eds), 56–67. Berlin: Springer-Verlag.

Clark, Roger A. 1978. "The transition from action to gesture." In *Action, Gesture and Symbol,* Andrew Lock (ed.), 231–257. London: Academic Press.

Dale, Philip S. and Fenson, Larry. 1996. "Lexical development norms for young children." *Behavioral Research Methods, Instruments, and Computers*, 28: 125–127.

Erikson, Erik. 1966. "Ontogeny of ritualization in man." *Philosophical Transactions of the Royal Society of London. Series B, Biological Sciences* 251: 337–349.

Genty, Emilie, Breuer, Thomas, Hobaiter, Catherine and Byrne, Richard W. 2009. "Gestural communication of the gorilla (*Gorilla gorilla*): Repertoire, intentionality and possible origins." *Animal Cognition* 12: 527–546.

Herbert, Jane, Gross, Julien and Hayne, Harlene. 2006. Age-related changes in deferred imitation between 6 and 9 months of age. *Infant Behavior and Development* 29: 136–139.

Iverson, Jana M., Capirci, Olga and Caselli, M. Cristina. 1994. "From communication to language in two modalities." *Cognitive Development* 9: 23–43.

Jakobson, Roman. 1962. "Why 'mama' and 'papa'?" In *Selected Writings, Vol. I: Phonological Studies* 538–545. The Hague: Mouton.

Lock, Andrew. 1978. "The emergence of language." In *Action, Gesture and Symbol,* Andrew Lock (ed.), 3–18. London: Academic Press.

Masur, Elise F. 1983. "Gestural development, dual-directional signaling, and the transition to words." *Journal of Psycholinguistic Research* 12: 93–108.

McNeill, David. 1998. "Speech and gesture integration." In *The nature and functions of gesture in children's communication. New directions for child development* Jana M. Iverson and Susan Goldin-Meadow (eds), 79: 11–27. San Francisco: Jossey-Bass Inc.

Messinger, Daniel S. and Fogel, Alan. 1998. "Give and take: The development of conventional infant gestures." *Merrill-Palmer Quarterly* 44: 566–590.

Piaget, Jean. 1962. *Play, Dreams, and Imitation in Childhood.* New York: Norton.

Pika, Simone, Liebal, Katja and Tomasello, Michael. 2003. "Gestural communication in young gorillas (*Gorilla gorilla*): Gestural repertoire, learning and use." *American Journal of Primatology* 60(3): 95–111.

Pika, Simone, Liebal, Katja and Tomasello, Michael. 2005. "Gestural communication in subadult bonobos (*Pan paniscus*): Gestural repertoire and use." *American Journal of Primatology* 65(1): 39–51.

Pika, Simone, Liebal, Katja, Call, Josep and Tomasello, Michael. 2005. "The gestural communication of apes." *Gesture* 5(1/2): 41–56.

Stokes, Trevor F., Baer, Donald M. and Jackson, Rachel L. 1974. "Programming the generalization of a greeting response in four retarded children." *Journal of Applied Behavior Analysis* 7: 599–610.

Tanner, Joanne E. and Byrne, Richard W. 1996. "Representation of action through iconic gestures in a captive lowland gorilla." *Current Anthropology* 37: 162–173.

Tomasello, Michael. 2008. *The Origins of Human Communication.* Cambridge, MA: MIT Press.

Tomasello, Michael. 2006. "Why don't apes point?" In *Roots of Human Sociality: Culture, Cognition, and Interaction* Nicholas J. Enfield and Stephen C. Levinson (eds), 506–524. Oxford: Berg.

Tomasello, Michael. 1996. "Do apes ape?" In *Social Learning in Animals: The Roots of Culture* Bennett J. Galef and Cecilia Heyes (eds), 319–346. San Diego, CA: Academic Press.

Tomasello, Michael, Call, Josep, Warren, Jennifer, Frost, G. Thomas, Carpenter, Malinda and Nagell, Katherine. 1997. "Ontogeny of chimpanzee gestural signals: A comparison across groups and generations." *Evolution of Communication* 1: 223–259.

Zinober, Brenda and Martlew, Margaret. 1985. "Developmental changes in four types of gesture in relation to acts and vocalizations from 10–21 months." *British Journal of Developmental Psychology* 3: 293–306.

Zukow-Goldring, Patricia and Arbib, Michael A. 2007. "Affordances, effectivities, and assisted imitation: Caregivers and the directing of attention." *Neurocomputing* 70: 2181–219.

A mother gorilla's variable use of touch to guide her infant

Insights into iconicity and the relationship between gesture and action

Marcus Perlman,[1,2] Joanne E. Tanner[2,3] & Barbara J. King[4]
[1]Department of Psychology, University of California, Santa Cruz / [2]The Gorilla Foundation / [3]School of Psychology, University of St. Andrews / [4]Department of Anthropology, College of William and Mary

This chapter examines how gestures of the great apes are created from instrumental actions. Ape gestures are generally believed to form through phylogenetic or ontogenetic ritualization, or – at least in humans – "iconic" gestures are created spontaneously during online interaction. These alternatives are evaluated with respect to data on the tactile *pushes* used by a mother gorilla to direct her infant around their enclosure. Analysis shows that the *pushes* exhibit variability in form and force in ways that are tuned to the present physical and social context, indicating the underlying activation of afforded instrumental actions and thus iconic processes in the creation of these gestures, opposed to ritualization. We discuss how this variability reveals continuity between gesture and action that is compatible with recent simulation-based accounts of iconic gesture.

1. Introduction

The great apes – human and non-human alike – use an array of bodily gestures to communicate with one another about action. Gestures may initiate play, grooming, and nursing, coordinate movement, indicate dominance and submission, and negotiate sexual interactions (Call & Tomasello 2007). A close relationship between gesture and action is to be expected (Arbib 2002; Armstrong & Wilcox 2007; King 2004; Tanner & Byrne 1996). Gesture, after all, is itself a sort of action, and there often exists a visible similarity between the form of a gesture and the implied action that constitutes the basis for its meaning. Indeed, researchers have examined the similarity between particular gestures and the actions they resemble to gain insight into the gestures' origins and the cognitive processes that enable their use.

Traditionally, however, there is thought to be an important psychological distinction between communicative acts, such as gestures, and direct instrumental acts on a social partner:

> [Communicative] acts achieve their ends indirectly. That is, the actor does not act physically to alter things to suit his needs, pushing or dragging other individuals about, beating them into submission, or the like. Instead, the actor's behavior provides other individuals with information, and the actions that *they* take on the basis of this information lead to any functions that are obtained. (Smith 1980: 389, quoted in Call & Tomasello 2007: 7)

Thus communicative gestures are considered to be those social acts that influence another individual by conveying information rather than by physical force, and it is typically assumed that these processes can be categorically distinguished from one another.

The processes by which a communicative gesture originates from an instrumental action are generally reasoned to occur along the time scales of phylogeny, ontogeny, or during the moments of online interaction. The first two of these are well accepted by ethologists to play a crucial role in the gestures used by both non-human and human apes (Eibl-Eibesfeldt 1972; van Hooff 1972). The third class of gestures, however, is more controversial. These gestures – often referred to as iconic – are created spontaneously and bear a visual or otherwise perceivable similarity to the meaning they express. Thus action-related iconic gestures are motivated by an active sense of an underlying instrumental act, such as Tomasello's (2008) creation of a sprinkling-grated-cheese gesture to place an order for Parmesan at an Italian cheese shop. While iconic gestures are believed to be a universal part of human communication (Kendon 2004; McNeill 1992), the ability of apes to use these gestures is debated (Call & Tomasello 2007; Pika 2007a,b; Tanner & Byrne 1996; Tanner, Patterson & Byrne 2006).

Our purpose in this chapter is to gain insight into the relationship between gesture and action by examining the *pushes* used by a gorilla mother to direct her infant around their enclosure. We first review research on the three basic ways that gestures are thought to arise from action, and then present description, analysis, and discussion of the mother-infant *pushes.*

1.1 Gestures from phylogenetic ritualization

The ethology literature offers many well-documented examples across a wide range of animal species of how a signal movement can evolve through the ritualization of a functional action sequence (Darwin 1872; Tinbergen 1952). This process – known as phylogenetic ritualization – is illustrated by the snarl of a wolf, a facial expression derived from the practical action of retracting the lips to bite. Another animal interacting with an ancestral wolf would have gained an advantage from reading this intention movement and anticipating the ensuing bite. In turn, a later generation of wolf that instinctively exaggerated this lip retraction could benefit by triggering other animals to respond as if to an oncoming bite, and so on through the generations to evolve the ritualized snarl observed today. This process is also believed to play a major role in the evolution of many of the bodily signals of the great apes including various facial expressions, postures, and vocalizations (Darwin 1872; Van Hooff 1972).

To the extent that ritualization is an operating force in the development of a gesture, it can be expected to display certain hallmark qualities related to its use and form (Blest 1961). One such quality is "emancipation" from the gesture's original proximal stimulus. The increasingly ritualized act, over the course of evolution, becomes independent of the causal factors that mediated its ancestral precursor, and is thereby free to become stylized into a standard, stereotyped form. Blest (1961) describes:

> Whereas stimuli of varying strength for the release of the unritualized precursors of display movements elicit responses of varying intensity and form, following ritualization the derived responses acquire an almost constant form and intensity to a wide range of stimulus strengths. (Blest 1961: 104)

However, the stereotyping forces of ritualization often compete with the advantage provided by the continuous variability exhibited by functional actions (Morris 1957). The result is often that a signal comes to be expressed with a stable, "typical intensity" in response to a wide but intermediate range of stimulus strengths, but can vary in response to more extreme stimuli. According to Morris, the variable expression of a ritualized signal is modulated essentially along two parameters – intensity and frequency – that correspond to differential motivation to display the signal. Yet, important to note is that a signal's variable intensity refers to the variable degree of enactment of the *ritualized* act. Thus, a maximally motivated ritualized signal manifests in an extreme form of the ritualized behavior, not the full execution of the instrumental act from which it was derived.

Specific accounts of how bodily signals of the great apes derive from functional actions through phylogenetic ritualization have predominantly been reserved for behaviors like facial expressions and have not often been offered for manual gestures (e.g. Eibl-Ebesfeldt 1972; Van Hooff 1972). However, one recent study takes a strong position consistent with the idea that phylogenetic ritualization is fundamental to the manual gestures of gorillas (Genty, Breuer, Hobaiter & Byrne 2009). Based on studies across three captive facilities and one wild population, Genty et al. identify a repertoire of 102 different gestures and conclude that it is built of species-typical, biologically inherited gestural forms. Although little explanation is offered for how these gestures would come to be innately specified, phylogenetic ritualization would presumably be a major operating force in their creation, as this is generally understood to be the main evolutionary process through which communicative signals are created (Morris 1957; Tinbergen 1952).

1.2 Gestures from ontogenetic ritualization

Researchers have also suggested that ape gestures develop over the course of an individual animal's interactional experience (Call & Tomasello 2007; Plooij 1978). Borrowing from the ethological concept, Tomasello and colleagues propose that a large number of great ape gestures arise through a process called ontogenetic ritualization. For example, a young chimpanzee may repeatedly raise his arm to play-hit a companion, to which his social partner gradually learns to anticipate that an arm-raise attends a play invitation (Tomasello 2008). The arm-raiser in turn comes to recognize his partner's anticipation and waits for a play response after making the arm-raise movement. Thus a ritualized gesture is created as two individuals anticipate and shape one another's behavior over repeated instances of the same interaction.

Importantly, by the essential nature of the ritualization process, ontogenetic ritualization also leads to stylized communicative gestures that lack graded variation in form and meaning (Burling 1999). Notably, however, this lack of variation is different from a lack of flexibility in the deployment of the gesture, which is granted by the emancipation of the signal from its original functional action.

1.3 Iconic gestures

Numerous researchers have also described apes' use of gestures derived more spontaneously from action. Tanner and Byrne (1996), for example, report the iconic gestures created by Kubie, a captive gorilla, who used them to coordinate playful action with a young female Zura. Using both visual and tactile gestures, Kubie indicated how he apparently desired Zura to move, with the direction of the gesture depicting the desired direction of movement. Kubie utilized a similar repertoire with three different females at different time periods, making ontogenetic ritualization an unlikely explanation. Tanner and Byrne (1999) also note a great deal of gradation and variation in the parameters of Kubie's iconic gestures that made them difficult to categorize as "types," a quality not predicted by ritualization.

A number of other studies also document the use of iconic gestures by apes (Crawford 1937; Hayes & Nissen 1971; Köhler 1925/1948; Pika & Mitani 2006; Russon & Andrews 2010; Savage-Rumbaugh et al. 1977; Savage-Rumbaugh et al. 1986; Tanner, Patterson & Byrne 2006; Yerkes 1943), with more sophisticated gestures described for enculturated apes. For instance, Kanzi and his sister Mulika, two bonobos raised with intensive human contact and symbolic communication from a young age, are reported to have spontaneously created iconic gestures for actions they wanted performed (Savage-Rumbaugh et al. 1986). Savage-Rumbaugh and colleagues describe that they "made twisting motions toward containers when they needed help in opening twist-top lids" and "hitting motions toward nuts they wanted others to crack for them" (p. 218).

2. Present study

Despite numerous reports of iconic gestures, studies on the origins of great ape gestures have tended to focus on individual learning processes such as ontogenetic ritualization (Call & Tomasello 2007) or biological inheritance (Genty et al. 2009), and have given less direct attention to iconic processes. Here we evaluate each of these three alternative explanations as they may account for the directive *pushes* used by a gorilla mother, Bawang, to guide her infant Barney.

Specifically, we examine (1) the variation in the form and physical force of Bawang's directive *pushes*; (2) whether this variation appears to be motivated by details of the physical and social context such as presently-afforded instrumental actions or Barney's attitude towards coordinating with his mother; and (3) whether Barney's attitude is influenced by the degree of physical force used by Bawang.

According to phylogenetic or ontogenetic ritualization, Bawang's directive *pushes* (if they function as communicative gestures), are stylized signals that ought to be consistent in form and distinct from any particular instrumental act, to which they would not be expected to exhibit any systematic, graded similarity. More or less similarity in the performance of the ritualized gesture to its original instrumental action would be random and not differentially meaningful to Barney. He would not, for example, be expected to respond more coordinatively to *pushes* that are performed with instrumental force, as these would be the same acts that are signaled by the ritualized gesture.

Alternatively, in an iconic account, Bawang's directive *pushes*, when communicative, are created spontaneously in the moment of interaction. They would be formed through an active sense of a particular instrumental action (i.e. actual physical manipulations of Barney), and thus may exhibit motivated variability in form, such as in angle and trajectory, as opposed to fixed stylization. Thus, an iconic gesture could potentially vary systematically in its degree of enactment and mechanical effectiveness, motivated, for instance, by the social context of Barney's attitude towards coordinating with his mother. In turn, Barney may be sensitive to this variation, responding more coordinatively to more forceful *pushes*.

3. Method

3.1 Subjects and setting

At the time of the video recording, the subjects, Bawang and her infant son Barney, were members of a stable social group at the San Francisco Zoo, San Francisco, USA. Bawang was 13 years old, and Barney, her second son, was 10–11 months old. The group also included Bawang's first son, five-year-old Shango, their father, the silverback Kubie, and two unrelated adult females Zura and Pogo. The gorillas had access to a large outdoor area (2,300 m^2, 38 x 50 m at maximum parameters) covered with grass and other vegetation and containing large, climbable live trees as well as several dead trees, large stumps, and rock structures.

3.2 Data recording and scene selection

Video recordings of Bawang and Barney were made once a week during August and September of 1994 by JET using focal animal sampling (Altmann 1974). Although JET's observations extended before and after this time period, it was only during this two-month window when Bawang was observed to use such directive *pushes*, and they often appeared motivated by Bawang's interest to remove Barney from situations of rough play with his older brother. JET selected all scenes recorded from this period in which Bawang used directive touches with Barney, which included in addition to *pushes*, various forms of *grabs, pulls, swings*, and *carries*. For the present work, we selected only clips of interactions involving two or more directive *pushes* within a contiguous interaction, resulting in 21 video clips. The mean length of each clip was 36.1 s, with a standard deviation of 11.9 s.

3.3 Coding and analysis

All directive *pushes* were analyzed, irrespective of mechanical effectiveness. Bawang's intent to direct Barney was established by observing that the two gorillas began moving immediately after the TOUCH, or when Barney did not begin moving, by Bawang's repeated efforts. The coding was conducted by MP, according to the operating definitions provided in Table 1. For reliability, a second coder, naïve to the hypotheses and theoretical motivation of the study, also coded the data according to the same criteria.

Pushes were considered as any act in which Bawang touched Barney with her palm or the back of her hand, without grabbing him, in an apparent attempt to direct him to move. Mechanical force for each was coded as *light, medium*, or *heavy/full*. Barney's attitude towards coordinating his movements with Bawang, both before and after each *push*, was coded as *passive* or *active* and *coordinative* or *discoordinative*.

Table 1. Definitions of coding categories

Force	**Amount of physical pressure applied by Bawang to Barney**
Heavy/Full	Substantial lasting physical impact on position and/or orientation, or transmits a sharply jarring impact.
Medium	Clear but not jarring physical impact on position and/or orientation; without a lasting, substantial change.
Light	Minimal physical effect on position and/or orientation.
Attitude	**Barney's attitude immediately before or after Bawang's *push***
Active	Actively moving or orienting.
Passive	Mostly still and passive.
Discoordinative	Acts in direct opposition to previous *push*, with either active or passive resistance.
Coordinative	Acts in accordance with previous *push* by moving in the desired direction; may be passive if not yet directed anywhere.
Form	**The form of Bawang's pushes**
Backhand	Pushes with the back of the hand.
Inside Out	Pushes with inside out motion with the palm of the hand away from the body.
Overhand	Pushes with a mostly forward-facing palm, may be directed downward.
Underhand	Pushes from underneath with an upward facing palm, sometimes rolling over the hand.

4. Results

The 21 qualifying sequences of Bawang and Barney contained a total of 54 *pushes* for analysis, with an average of 2.57 *pushes* per sequence and standard deviation of 1.72 (four instances were discarded because physical contact could not be clearly determined). Table 2 summarizes the use of force by each type of *push* and also by Barney's attitude immediately prior to contact. For comparison, the results from both coders are presented.

Table 2. Use of force by *push* type and Barney's attitude

	Total	Light	Medium	Heavy/Full
***Push* Type**				
Backhand	6 (11.1%) 8 (14.8%)	2 (33.3%) 3 (37.5%)	3 (50.0%) 5 (62.5%)	1 (16.7%) 0 (0.0%)
Inside out	2 (3.7%) 2 (3.7%)	2 (100.0%) 2 (100.0%)	0 (0.0%) 0 (0.0%)	0 (0.0%) 0 (0.0%)
Overhand	35 (64.8%) 33 (61.1%)	15 (42.9%) 11 (33.3%)	9 (25.7%) 13 (39.4%)	11 (31.4%) 9 (27.3%)
Underhand	11 (20.4%) 11 (20.4%)	3 (27.3%) 3 (27.3%)	3 (27.3%) 3 (27.3%)	5 (45.4%) 5 (45.4%)
Total	54 54	22 (40.7%) 19 (35.2%)	15 (27.8%) 21 (38.9%)	17 (31.5%) 14 (25.9%)

(*Continued*)

Table 2. Use of force by *push* type and Barney's attitude (Continued)

	Total	Light	Medium	Heavy/Full
Prior Attitude				
Coordinative–	23 (43.4%)	13 (56.5%)	7 (30.4%)	3 (13.0%)
Active	22 (41.5%)	12 (54.5%)	6 (27.2%)	4 (18.2%)
Coordinative–	4 (7.5%)	3 (75.0%)	0 (0.0%)	1 (25.0%)
Passive	10 (18.9%)	5 (50.0%)	4 (40.0%)	1 (10.0%)
Discoordinative–	9 (17.0%)	3 (33.3%)	3 (33.3%)	3 (33.3%)
Active	7 (13.2%)	1 (14.3%)	5 (71.4%)	1 (14.3%)
Discoordinative–	17 (32.1%)	3 (17.6%)	5 (29.4%)	9 (53.0%)
Passive	14 (26.4%)	1 (7.1%)	6 (42.9%)	7 (50.0%)

In each cell: Top two numbers from Coder 1, bottom two numbers from Coder 2; left number = number of *pushes*, right number = percentage of *pushes* by force for given *push* type or attitude (three right columns) or percentage of total for given *push* type or attitude (left column).

4.1 Form and force

Four basic types of *pushes* were identified (see Table 1) and were confirmed by the second coder with 85% reliability (Cohen's kappa k = 0.73). Overhand *pushes* were used most frequently (n = 35, 64.8%), but underhand (n = 11, 20.4%) and backhand *pushes* (n = 8, 11.1%) were also used with substantial frequency. Inside out *pushes* were used just twice (3.7%).

The distribution of force exhibits a considerable amount of variation across *push* types (coded with 71.7% agreement, linear Cohen's kappa k = 0.67; see Table 2). Excluding the rare inside out form, each other type of *push* was employed with light, medium, and heavy/full force, with each category represented at least 16.7% of the time. Overall, 40.7% of the observed *pushes* were performed with light, non-instrumental force, 27.8% with medium force and 31.5% with heavy/full force.

4.2 Attitude

Barney's attitude was coded immediately before and after Bawang performed each *push*. First it was determined whether his willingness to coordinate with his mother influenced the degree of force she used. Barney's attitude was coded by the second rater with 83.3% agreement for active/passive (k = 0.66) and 84.9% agreement for coordinative/discoordinative (k = 0.69).

A chi-square compared the distribution of force for Barney's coordinative and discoordinative attitudes, showing a significant difference between them ($\chi^2 = 8.60$, df = 2, $p < 0.05$; for the second coder, $\chi^2 = 10.76$, df = 2, $p < 0.01$). Specifically, Barney's discoordinative attitude was associated with more forceful *pushes* by Bawang.

We also looked at Barney's attitude immediately following Bawang's *pushes* to determine whether it was influenced by the degree of force used by Bawang. Barney's subsequent attitude was coded by the second rater with 88.9% agreement for active/passive (k = 0.74) and 87.0% agreement for coordinative/discoordinative (k = 0.72). Of 27 interactions in which Barney began with a coordinative attitude, his attitude changed to discoordinative only twice, once after a light *push* and once after a medium *push*. Conversely, of 26 interactions in which Barney began with a discoordinative attitude, his attitude changed to coordinative five times, four times after a heavy/full *push* (4 of 12 heavy/full *pushes* or 33.3% led to reversal) and once after a medium *push* (1 of 14 light and medium *pushes* or 7.1% led to reversal). The small number of observations in these conditions was not sufficient to reach statistical significance ($\chi^2 = 1.42$, df = 1, p = 0.23).

5. Discussion

The present study examined the *pushes* used by a gorilla mother, Bawang, to direct her infant Barney around their enclosure. Specifically, it sought to assess the variation (or lack of it) in the form and physical force of Bawang's directive *pushes* and to determine whether it appeared to be motivated by details of the physical and social context. We further examined whether variation in the physical force of Bawang's *pushes* might be differentially meaningful to Barney, influencing his willingness to coordinate action with his mother. Our goal was to gain insight into how gestures may originate from instrumental actions by comparing an iconic account of gesture to theories of ontogenetic and phylogenetic ritualization in their ability to explain the results of the study.

The results show that Bawang's *pushes* are variable along several dimensions. They frequently vary in their degree of mechanical effectiveness, sometimes employing an instrumental degree of force, other times a light, ineffective TOUCH, and yet other times a degree of force somewhere in between. Furthermore, they appear molded to the context at hand and reflect, at least in a number of cases, actually afforded instrumental manipulations of Barney. For example, Bawang tended to employ directive touches involving an upward facing palm and lifting movement, an underhand *push,* when Barney was passively lying or sitting down. Of eleven total underhand *pushes*, eight (72.7%) were used when Barney was passive, compared to eleven of 35 overhand *pushes* (31.4%) used in this context.

Bawang's *pushes* also appear to be tuned to the social context of Barney's attitude, a point demonstrated by her tendency to apply more forceful, and hence, more "instrumental" *pushes* when Barney was being discoordinative, and less forceful, more "gestural" *pushes* when he was coordinative. In turn, there is tentative evidence (limited by the number of observations) to suggest that Barney is responsive to the differential use of force by his mother, showing an increased tendency to change from a discoordinative to coordinative attitude after the application of heavy/full forced *pushes*.

This physically and socially motivated variability in Bawang's *pushes* is not consistent with the hypothesis that ritualization, whether through evolution or ontogeny, is predominantly responsible for their creation. Both processes of ritualization lead to stylized forms that are consistently distinct from any particular afforded instrumental act. Instead, Bawang's directive *pushes* appear actively shaped by the physical and social context, with even her less forceful gestures, through their orientation and trajectory, formed according to the afforded *push* that would move Barney in the desired direction. Moreover, Barney appears able to understand his mother's intentions (e.g. her desired direction, her emotional state) through the particulars of the *push* as it resembles an actual instrumental act. These observations all point to an active role of iconic processes in the creation of Bawang's *pushes*.

One potential reservation in interpreting these results rests in the reliability of the data coding. Variables like degree of force and attitude are difficult to determine precisely, and this imprecision is reflected in the moderate degrees of inter-rater reliability. However, sufficient reliability exists to confirm a substantial degree of variability in the force and form of the *pushes*, and as evident from Table 1, the two coders observed very similar patterns of variability. Moreover, some limited degree of variance is expected in the categorization of continuously varying as opposed to categorically distinct forms of behavior such as ritualized gestures.

5.1 Iconicity and gesture-action continuity

Reports observing the ability of the great apes to spontaneously create iconic gestures have often been received with a good deal of skepticism (Call & Tomasello 2007; Pika 2007a, 2007; Tomasello 2008). Consider for example, the iconic gestures reported by Tanner and Byrne (1996) in which a silverback gorilla (Kubie) moved his arms in a particular direction to communicate to a female (Zura) to move in that direction (i.e. gestures quite similar to the ones reported here). According to Tomasello (2008), these are most likely "garden-variety ritualized behaviors" that appear iconic to human observers because they "derive from attempts to actually move the body of the other in the desired direction" (2008:27). Such dismissal is summarized by Tomasello and Call's conclusion that "Iconic gestures... are basically not used by apes" (2007:234).

We suggest that a primary reason for doubting that apes are able to produce iconic gestures is based on the *a priori* assumption that the use of iconic gestures depends on highly developed social-cognitive abilities related to imitation and theory of mind. Tomasello, for instance, reasons (2008: 203):

> To use an iconic gesture one must first be able to enact actions in simulated form, outside their normal instrumental context – which would seem to require skills of imitation, if not pretense... [T]o comprehend an iconic action as a communicative gesture, one must first understand to some degree the Gricean communicative intention; otherwise the recipient will suppose that the communicator is simply acting bizarrely, trying to run like an antelope or to dig a hole for real when the context is clearly not appropriate.

Tomasello and colleagues argue that the great apes have only limited abilities to imitate and share communicative intentions, and therefore are unable to use iconic gestures (Call & Tomasello 2007; Tomasello 2008).

Yet, like the directive *pushes* described in the present study, reports of iconic gestures by the great apes generally describe them as occurring fairly directly *within* "their normal instrumental context." Because these gestures are thus contextualized, they may be comprehended without necessarily understanding the gesturer's "Gricean communicative intention," but directly through an activated sense of the instrumental action within a context that is relevant to exactly that sort of action. These gestures need not be "quarantined" because they are not – in production or comprehension – fully distinct from the action for which they are iconically expressive.

In addition to work by Tomasello, Call, and colleagues, a comprehensive study by Genty et al. (2009) presumably also failed to observe iconic gestures, although they do not explicitly report one way or the other. Instead, the authors identify 102 different gesture types across several wild and captive groups of gorillas, and based on the consistent observation of overlap, they suggest that the gestures must each be part of a biologically inherited repertoire. One may note, however, that 102 different types of gestures is a lot, and perhaps as a necessary constraint on such a large number, many of the categories seem to include variations on similar kinds of physical motions. For example, several of the gestures (*2-handed grab, 2-handed grab-pull, grab-pull, one-handed grab, positioning, hand on, hands on, 1-handed push, 2-handed push, touch*) involve touching some unspecified part of another with the palm of the hand and bear a notable similarity to the *pushes* described in the present study. (In addition to *pushes*, Bawang also directed Barney with a variety of *pulls*, *swings*, and *grabs* that were not reported here.) Some of these gestures may have been split by the authors into multiple categories when they just as reasonably could have been included together in a single category, thus indicating a certain amount of arbitrariness to the categorical designations.

Indeed, the 102 different gesture types appear to fill a substantial portion of the full range of actions that are anatomically possible for a gorilla to perform. Especially as many similar gestures were observed in several but not all the populations studied, the overlap could instead be a result of probable action responses to recurring affordances common to gorilla environments and social lives. Further research might distinguish whether these gesture types are truly separate categories and are thus likely to be ritualized, or whether they instead reflect variations on commonly afforded actions, which would lend support to an iconic account.

Whereas most previous work makes a categorical distinction between gesture and instrumental action, the nature of the directive *pushes* described in this study points to a more continuous relationship between them in which the distinction is more a matter of degree. *Push*-like gestures may be created and understood through an activated sense of the full instrumental act (such as a hard, sustained *push*) that is expressed through a weakened or partially-enacted version (a gentle TOUCH). Bawang may have been motivated to cause Barney to move in a particular direction by means of an instrumental act upon him, but, perhaps for socially-minded considerations, she was inclined to inhibit the full act and carry it out in a weakened or partially-enacted form. At other times, particularly when Barney was being less coordinative, Bawang may have become emotionally aroused, leading to less inhibition of the instrumental act and increased application of force upon Barney.

The literature on great ape iconic gesture suggests a variety of social considerations that might lead to the inhibition of instrumental acts. In the present study, for example, Bawang may have had an interest in fostering Barney's locomotive independence, as it would be quite exhausting to have to *push* him everywhere she wanted him to go. Thus she would be motivated to communicate her intentions rather than forcefully carry out the full instrumental act.

In the case of Tanner and Byrne's (1996) report of Kubie and Zura, they note how special circumstances – Zura's ability to escape because of her small size and the intimidating presence of a more dominant silverback – pressured Kubie to be especially tactful and convincing if he wanted to engage Zura in playful interactions. If he were too forceful and unpleasant, Zura could easily have put an end to the bout. And in the case of the iconic gestures produced by Kanzi and Mulika (Savage-Rumbaugh et al. 1986), the inhibition appears to have been for physical, not social reasons. The desired items – nuts and a jar with a twist-top lid – were simply out of physical reach, and so the instrumental action could not be performed. The result in each of these cases is that a gesture is born out of the partial enactment of an instrumental act that is restrained for one reason or another.

5.2 Gesture and sensorimotor simulations

The puzzle remains of how to get from rather simple iconic *pushes* to the full-blown iconic gestures that are considered characteristically human, such as Tomasello's (2008) cheese-sprinkling gesture. How could the so-called iconic gestures of apes be at the origin of human gesture? One possible explanation begins by considering the compatibility of the observed continuity between gesture and action with recent accounts that hypothesize sensorimotor simulations as the essential cognitive process involved in the production of iconic gestures. According to one such view, simulations of action-related thoughts lead to the activation of neural premotor action states, which then has the potential to spread to motor areas (Hostetter & Alibali 2008). This spreading activation comes to be realized as the overt action of iconic gesture.

The idea that iconic gestures are rooted in the simulation of actual instrumental acts is demonstrated in an experiment by Cook and Tanenhaus (2009) in which participants solved the Tower of Hanoi problem and described their solution to a listener. In this problem, a stack of disks is arranged bottom-up from largest to smallest on the leftmost of three pegs, and the goal is to move all of the disks to the rightmost peg, moving only one disk at a time and without ever placing a larger disk on top of a smaller one. Participants performed the task either by physically manipulating a real stack of weights or with a mouse in a computer version of the problem.

When participants described their solution, their iconic gestures were found to correspond to the actual trajectory involved in solving the problem, with differences reflecting the differently afforded constraints of the real-weight versus computer task. Cook and Tanenhaus interpret these analog differences as evidence for the activation of perceptual-motor information that is involved in the actual performance of the task. The instrumental act is simulated, leading to the creation of an iconic gesture that is a direct reflection of this activation.

What are the implications of applying a simulation-based view to the iconic gestures of the great apes? Perlman and Gibbs (in press) consider this question and pose some predictions about the characteristics and qualities of the gestures one would expect to find by the hypothesis. They suggest that the most phylogenetically primitive iconic gestures would be those that manifest from the most imaginatively simple of simulations. Thus, for example, the simulation ought to involve the gesturer's own perspective and body, and not someone else's, let alone the gesturer's body imagined as something entirely different. In addition, the contextual and afforded elements of the simulation ought to be largely present in the gesturer's immediate perceptual experience – the more distant the element, presumably the more difficult it would be to imagine. In sum, primitive iconic gestures ought to be tightly connected – in form, meaning, and context – to the presently afforded instrumental and attentive actions that are available to the gesturer. And indeed, as we have seen, the most commonly observed iconic gestures are those that are most concrete in form and function, such as tactile gestures used to direct an interlocutor's movement.

Moreover, there is evidence that with increased human enculturation and immersion in language routines, there is an expansion in apes' imaginative ability to engage in mental simulations and mime the imagined actions. For instance, recall the iconic gestures produced by Kanzi and Mulika, notable in how they incorporate the imagined physical manipulation of objects that are not available to immediate tactile experience (Savage-Rumbaugh et al. 1986). Yet, as evident from their description, the simulated objects *are* available to their visual perception. As Kanzi and Mulika perform these gestures, their visual attention is clearly drawn towards the desired objects. One might wonder whether they could produce such sophisticated iconic gestures with the same facility if the objects were located outside of their perceptual purview.

According to the simulation hypothesis, the capacity to create iconic gestures could be considered along the degree of complexity involved in the corresponding mental simulation. For example, a simulated transitive act in which the resulting gesture is directly enacted upon a present object (e.g. Bawang's *pushes* upon Barney) would require less simulative complexity than a simulated act in which the gesture is acted upon an imagined object (e.g. Tomasello's sprinkling-cheese gesture). Indeed, researchers have previously suggested developmental sequences of action-derived gestures through phylogeny and ontogeny that are compatible with the above ideas (Arbib 2002), and similar notions have been applied to apes without enculturation (Savage-Rumbaugh et al. 1977; Tanner et al. 2006) as well as to enculturated apes who show an accelerated sequence (Tanner et al. 2006). Thus, the bridge between Bawang's iconic directive *pushes* and more prototypically human-like iconic gestures may be explained through the development of an increasingly sophisticated ability to engage in sensorimotor simulations.

6. Conclusion

We have made the case that the processes of phylogenetic and ontogenetic ritualization are inadequate as comprehensive explanations for how ape gestures come to be formed from instrumental actions. The directive *pushes* that we have described appear, to a significant extent, to be molded to the physical affordances and social context of the moment of communication and thus reveal the online activation of a particular instrumental action. We offer that multiple processes are involved in the creation and use of ape gestures, including those operating along different time scales like phylogenetic and ontogenetic ritualization, but critically, one of those processes is the on-line adaptation of action to form spontaneous iconic gestures.

Our focus on gestures within the context of the full scope of more-or-less forceful actions has led to a markedly different conception than commonly assumed of the nature of iconic gestures and the cognitive processes that their creation and use entails. In particular we highlight the consideration of mechanical effectiveness as a variable along a spectrum from action to gesture. In a simulation-oriented view, there is not necessarily a sharp distinction between an iconic gesture and the instrumental act it represents, and accordingly, our data have revealed a notable amount of continuity between Bawang's gestural and instrumental *pushes*.

At the same time, while gesture-action continuity may lie at the origin of ape iconic gesture, it is crucial to consider how iconic gestures may come to be increasingly distinct from contextual elements involved in the real instrumental act. The greatly expanded capability to abstract gestures from their instrumental context is, we suggest, a fundamental cognitive skill that enables the human propensity for iconic gesture. One exciting path for future research is the further investigation of the conceptual and simulative qualities exhibited by the iconic gestures of enculturated and language-taught apes. By examining how these qualities increase in sophistication, we are likely to gain new insights into the nature of human gesture and imagination.

Acknowledgments

We are grateful for the very helpful comments and suggestions of the editors Simone Pika and Katja Liebal and two anonymous reviewers. We would also like to express our great appreciation to Jenny Hoffseth for her assistance as a second coder.

References

Altmann, Jeanne. 1974. "Observational study of behaviour: Sampling methods." *Behaviour* 49: 227–267.

Arbib, Michael A. 2002. "The mirror system, imitation, and the evolution of language." In *Imitation in animals and artifacts*, Chrystopher Nehaniv and Kerstin Dautenhahn (eds), 229–280. Cambridge, MA: The MIT Press.

Armstrong, David F. and Wilcox, Sherman E. 2007. *The Gestural Origin of Language*. New York: Oxford University Press.

Blest, A. David. 1961. "The concept of ritualization." In *Current Problems in Animal Behaviour*, William H. Thorpe and Oliver L. Zangwill (eds), 102–124. Cambridge: At the University Press. (Original work published 1961)

Burling, Robbins. 1999. *The Talking Ape: How language evolved.* Oxford: Oxford University Press.

Call, Josep and Tomasello, Michael. 2007. *The Gestural Communication of Apes and Monkeys.* New York: Lawrence Erlbaum Associates.

Cook, Susan W. and Tanenhaus, Michael. K. 2009. "Embodied communication: Speakers' gestures affect listeners' actions." *Cognition* 113:98–104.

Crawford, Meredith P. 1937. "The cooperative solving of problems by young chimpanzees." *Comparative Psychology Monograph* 14:1–88.

Darwin, Charles. R. 1872. *The Expression of the Emotions in Man and Animals.* London: Murray.

Eibl-Eibesfeldt, Irenäus. 1972. "Similarities and differences between cultures in expressive movements." In *Non-Verbal Communication,* Robert A. Hinde (ed.), 209–237. Cambridge: Cambridge University Press.

Genty, Emilie, Breuer, Thomas, Hobaiter, Catherine and Byrne, Richard W. 2009. "Gestural communication of the gorilla (*Gorilla gorilla*): Repertoire, intentionality, and possible origins." *Animal Cognition* 12:527–546.

Hayes, Keith J. and Nissen, Catherine H. 1971. "Higher mental functions of a home-raised chimpanzee." In *Behavior of Nonhuman Primates: Modern Research Trends,* Allan M. Schrier and Fred Stollnitz (eds), 59–115. New York: Academic Press.

Hostetter, Autumn B. and Alibali, Martha W. 2008. "Visible embodiment: Gestures as simulated action." *Psychonomic Bulletin & Review* 15: 495–514.

Kendon, Adam. 2004. *Gesture: Visible action as utterance.* Cambridge: Cambridge University Press.

King, Barbara J. 2004. *The Dynamic Dance: Nonvocal communication in African great apes.* Cambridge, MA: Harvard University Press.

Köhler, Wolfgang. 1948. *The Mentality of Apes.* London: Routledge and Kegan Paul. (Original work published 1925)

McNeill, David. 1992. *Hand and mind: What gestures reveal about thought.* Chicago: Chicago University Press.

Morris, Desmond. 1957. "'Typical intensity' and its relation to the problem of ritualization." *Behaviour* 11:1–12.

Perlman, Marcus and Gibbs, Raymond. W. Jr. in press. "Sensorimotor simulation in speaking, gesturing, and understanding." In *Body-Language-Communication: An international handbook,* Cornelia Müller, Ellen Fricke, Alan Cienki and David McNeill (eds). Berlin: Mouton.

Pika, Simone. 2007a. "Gestures in subadult bonobos (*Pan gorilla*)." In *The Gestural Communication of Apes and Monkeys,* Josep Call and Michael Tomasello (eds), 41–67. New York: Lawrence Erlbaum Associates.

Pika, Simone. 2007b. "Gestures in subadult gorillas (*Gorilla gorilla*)." In *The Gestural Communication of Apes and Monkeys,* Josep Call and Michael Tomasello (eds), 99–130. New York: Lawrence Erlbaum Associates.

Pika Simone and Mitani, John. C. 2006. "Referential gestural communication in wild chimpanzees (*Pan troglodytes*)." *Current Biology* 16:R191–R192.

Plooij, Frans. X. 1978. "Some basic traits of language in wild chimpanzees?" In *Action, Gesture and Symbol,* Andrew Lock (ed.), 111–131. London: Academic Press.

Russon, Anne and Andrews, Kristin. 2010. "Orangutan pantomime: Elaborating the message." *Biology Letters* 7(4):627–630.

Savage-Rumbaugh, E. Sue, Wilkerson, B. J. and Bakeman, R. 1977. "Spontaneous gestural communication among conspecifics in the pygmy chimpanzee (*Pan paniscus*)." In *Progress in Ape Research,* Geoffrey H. Bourne (ed.), 97–116. New York: Academic Press Inc.

Savage-Rumbaugh, E. Sue, McDonald, Kelly, Sevcik, Rose A., Hopkins, William D. and Rupert, Elizabeth. 1986. "Spontaneous symbol acquisition and communicative use by pygmy chimpanzees (*Pan paniscus*)." *Journal of Experimental Psychology: General* 115: 211–235.

Smith, William J. 1980. *The behavior of communicating: An ethological approach.* Cambridge: Harvard University Press.

Tanner, Joanne E. and Byrne, Richard W. 1996. "Representation of action through iconic gesture in a captive lowland gorilla." *Current Anthropology* 37: 162–173.

Tanner, Joanne E. and Byrne, Richard W. 1999. "The development of spontaneous gestural communication in a group of zoo-living lowland gorillas." In *The Mentalities of Gorillas and Orangutans: Comparative Perspectives,* Sue T. Parker, Robert W. Mitchell and H. Lyn Miles (eds), 211–239. Cambridge, UK: Cambridge University Press.

Tanner, Joanne E. and Byrne, Richard W. 2010. "Triadic and collaborative play by gorillas in social games with objects." *Animal Cognition* 13: 591–607.

Tanner, Joanne E., Patterson, Francine G. and Byrne, Richard W. 2006. "The development of spontaneous gestures in zoo-living gorillas and sign-taught gorillas: From action and location to object representation." *Journal of Developmental Processes* 1: 69–102.

Tinbergen, Nikolaas. 1952. "'Derived' activities: Their causation, biological significance, origin, and emancipation during evolution." *The Quarterly Review of Biology* 27: 1–32.

Tomasello, Michael. 2008. *Origins of Human Communication.* Cambridge: MIT Press.

Van Hooff, Jan A. R. A. M. 1972. "A comparative approach to the phylogeny of laughter and smiling." In *Non-Verbal Communication,* Robert A. Hinde (ed.), 209–237. Cambridge: Cambridge University Press.

Yerkes, Robert M. 1943. *Chimpanzees: A Laboratory Colony.* New Haven: Yale University Press.

Spontaneous use of gesture sequences in orangutans

A case for strategy?

Sebastian Tempelmann[1,2] & Katja Liebal[1,2,3]
[1]Department of Developmental and Comparative Psychology, Max-Planck Institute for Evolutionary Anthropology, Leipzig, Germany / [2]Department of Evolutionary Psychology, Freie Universität Berlin, Berlin, Germany / [3]Department of Psychology, University of Portsmouth, Portsmouth, UK

Great apes use gestures flexibly across different contexts and with a considerable degree of individual variability. However, little is still known about whether great apes combine gestures to increase the efficacy of their communicative attempts – either by anticipatorily attracting the attention of the recipient towards themselves or by continuing to gesture in case the recipient does not respond. The few existing studies are not consistent in their conclusions. The present study thus aims to contribute to this discussion by providing a systematic investigation of gesture sequences of 16 captive orangutans.

Our results showed that 20% of their gestures were part of a sequence and consisted mostly of combinations of two gestures with the majority representing repetitions of the same gesture. In addition, senders continued to gesture regardless of whether recipients responded to the initial gesture, they did not use particular gestures at the beginning of a sequence to attract the attention of recipients, and they did not switch to more efficient gestures in case the recipient did not respond. These findings therefore do not support the assumption that orangutans' use of gesture sequences reflects goal directedness in achieving a particular goal, but may rather be the result of the arousal of the gesturing individual in particular contexts, such as play.

1. Introduction

Gestures are an important part of human communication. Before children start to speak, they use a variety of gestures (e.g. Bates, Camaioni & Volterra 1975) and even after they have acquired language, gestures are still closely intertwined with the spoken modality (Kendon 2004). Given the role of gestures in human communication, different evolutionary scenarios have been proposed emphasizing the gestural origin of human language (e.g. Corballis 2002; King 2004). As a consequence, an increasing number of studies used a comparative approach to investigate the gestural communication in great apes and other non-human primate species (see e.g. Call & Tomasello 2007 for an overview). By looking at our closest living relatives, we are able to identify similarities but also differences in gesture use between non-human and human primates and based on this, to propose possible scenarios of the evolution of human communication. In this framework, an important issue is whether communication is used *intentional* in the sense of a voluntary controlled signaling, directed at a particular recipient to influence his behavior and thus to achieve particular goals. Different criteria have been used to distinguish intentional communication from an innate and/or conditioned stimulus-response signaling (e.g. Bates et al. 1975; Call & Tomasello 2007), including (1) the flexible use of communicative means, (2) the assessment of the recipient's attentional state, and (3) the goal directedness of the communicative act. In the following, we will briefly introduce these issues together with the current state of research regarding apes' gestural communication.

First, the flexible use of gestures indicates ontogenetic learning processes, thus arguing against genetically determined behavior (but see e.g. Genty, Breuer, Hobaiter & Byrne 2009 for a different conclusion). Apes use gestures in very flexible ways; that is, one gesture can be used to achieve different social goals or vice versa, several gestures can be used to obtain a particular social goal (means-ends dissociation, Bruner 1981; Pika & Liebal this volume). There is also a considerable degree of variability between individual gestural repertoires, both within and between groups, partly because individuals can invent and incorporate new gestures into their repertoires over lifetime (e.g. for bonobos: Pika, Liebal & Tomasello 2005; for chimpanzees: Tomasello, Call, Warren, Frost, Carpenter & Nagell 1997; for gorillas: Pika, Liebal & Tomasello 2003; for orangutans: Liebal, Pika & Tomasello 2006).

Second, the senders' capability to adjust their gesture use to the behavior of the audience is often measured in terms of the kind of gestures utilized depending on the attentional state of the recipient. In fact, all great ape species are sensitive to the attentional state of their communicative partners, both in interactions with conspecifics and with human experimenters (e.g. Call & Tomasello 2007; Hostetter, Cantero & Hopkins 2001). Here, all great ape species use the orientation of the face and not the body to assess the other's attentional state (Tempelmann, Kaminski & Liebal 2011).

Third, the sender's persistence to the goal is another feature used to characterize the intentional use of signals (e.g. Call & Tomasello 2007; Pika & Liebal this volume). It is usually measured by the sender's persistence to communicate when the previous communicative act failed, that means to repeat or to elaborate the communicative means (and to terminate the interaction when the goal has been achieved) (e.g. Bates et al. 1975). In addition to those three criteria, the sender's waiting for the recipient's response (Tomasello et al. 1997) and the use of attention-getting signals (e.g. Bates et al. 1979; Leavens, Russel & Hopkins 2005) to attract the recipient's attention for a following communicative act, are used to identify instances of intentional communication.

When focusing on gestural communication, both persistence to the goal and the use of attention getters (when the respective gesture is used to direct the attention of a potential recipient towards a following gesture) result in the emergence of gesture sequences. In other words, the use of gesture sequences enables an individual to (1) use attention getting behaviors in order to draw the attention of the potential recipient to the signaler and a subsequently produced gesture, and to (2) persist if the first communicative attempt fails, which becomes evident by the repetition or elaboration of the communicative act (e.g. Bates et al. 1979).

Therefore, the question is whether great apes combine gestures in a way that such combinations serve particular functions that single gestures do not. In other words, do gesture combinations provide a tool to increase the efficacy of great apes' gestural communication? The systematic examination of gesture sequences therefore can provide more detailed insight into the cognitive skills of great apes underlying their gestural communication. In the following paragraph, we will give a brief overview of existing studies on gesture sequences in apes with a focus on their persistence in communication and their use of attention getters.

Chimpanzees are reported to combine attention-getting gestures such as *ground slap* or *poke at* with visual gestures to make sure that a given conspecific will perceive the latter (Tomasello, Call, Nagell, Olguin & Carpenter 1994). However, other studies on chimpanzees (Liebal, Call & Tomasello 2004a) but also gorillas (Genty & Byrne 2010) did not find any evidence for a tactical use of attention-getting gestures as first element of a sequence in order to manipulate the attentional state of the recipient. However, these studies showed that individuals of both species use other strategies to make sure that their gestures will be perceived by recipients (see also Tanner 2004). Concerning experimental evidence, Liebal and colleagues showed that all four great ape species preferable move into the visual field of recipients to beg for food than using attention getters (Liebal, Pika, Call & Tomasello 2004b). Interestingly, orangutans and also gorillas – as opposed to chimpanzees and bonobos – were less likely to move in front of the human if the food was at a different location.

Concerning persistence, results of the few existing studies are inconsistent. For instance, both captive and wild chimpanzees repeat a gesture if the first failed to reach their communicative goal, but they do not systematically modify their gestures in response to the recipients' behavior (Liebal et al. 2004a; Hobaiter & Byrne 2011). In addition, Hobaiter and Byrne (2011) suggested that some types of gesture sequences reflect the chimpanzees' persistence in their communicative attempts, while others (*rapid-fire* sequences with less than one second in between) are used for inter-individual behavioral adjustment. In gorillas, gesture sequences did not emerge as result of the recipient's lack of response, but similar to chimpanzees they are used for inter-individual adjustment (Genty & Byrne 2010; Hobaiter & Byrne 2011). For orangutans, Cartmill and Byrne (2010) found that captive individuals indeed take into account the behavior of the recipient to an initial gesture and accordingly adjust their subsequent gesture use.

In summary, observational studies indicate that only chimpanzees' use of gesture sequences is characterized by their persistence in terms of repeating a gesture when a previous one did not reach the communicative goal. However, experimental studies investigating interspecies human-ape communication produced different results. Both, chimpanzees and orangutans, repeated their gestures when the human experimenter did not offer the full amount of food they begged for. In addition, they even elaborated their gesture use when the human did not offer the correct piece of food, and the same was found for orangutans (Cartmill & Byrne 2007; Leavens et al. 2005).

Thus, great apes seem to use different strategies in their interactions with conspecifics and humans. Although those experimental studies impressively demonstrate what apes can potentially do when raised in a specific environment, at the same time their interactions with humans are constrained by the experimental setting forcing them to use other forms of communication different from how they interact with conspecifics. Thus, to reconstruct the evolutionary pathway of gestural communication and to address questions regarding the actual function and relevance of gestural communication in a species, it is necessary to examine the natural communication between conspecifics. Although there is a steady increase of studies investigating the gestural communication in great apes in both captive and wild settings, we still know very little about gestural communication of orangutans and we know virtually nothing about their use of gesture sequences; that is, whether they are the result of communicative persistence and/or whether they are used strategically to influence the behavior of others. Since there are only few studies investigating the emergence of gesture sequences in great apes in general and the existing findings are not even consistent, it is important to systematically study a variety of different species. Since the form of social organization seems to have an impact on the complexity of gestural communication in non-human primates (Call & Tomasello 2007; Maestripieri 1999), orangutans are especially interesting in this context, as their lifestyle is different from those of the other great apes (e.g. van Schaik 1999). Orangutans are a semi-solitary species and their individual-based fission-fusion system (Delgado & van Schaik 2000) is different from bonobos and chimpanzees living in large fission-fusion groups, but also from the gorillas' small and stable family groups (Smuts et al. 1987).

The present study is the first systematic investigation of gesture sequences in captive orangutans focusing on interactions with conspecifics, therefore filling a gap in research on great apes' gestural communication. The aims of the present study were to (1) identify and describe the gesture types used in gesture sequences and the contextual background of their usage, (2) investigate gesture repetitions and elaboration as a function of the recipient's behavior, and (3) examine whether so called attention getters (that are gestures which are appropriate to manipulate the attentional state of a recipient) are used strategically to influence the behavior of recipients. In this context it is important to note that attention getters may be used in two different ways: Either they are used as initial signal in a sequence and may represent an anticipation of a communicative interaction, or they occur after an initial gesture as respnse to a non-responding recipient, indicating the sender's persistence and elaboration.

In order to address our first research aim, we will describe the kinds and frequency of gesture combinations used by orangutans and the functional context they occur in to investigate whether orangutans' use of gesture sequences is different from that of single gestures. To investigate our second aim, we will focus on the recipient's reaction to the initial gesture of a sequence. We predict that when recipient's respond less often to initial gestures compared to single gestures, gesture sequences very likely emerge because of the senders' persistence in their communicative attempts. Furthermore, we will analyze whether orangutans repeat their initial gesture or switch to another gesture indicating an elaborated gesture use as a function of the recipients' reaction. To address the third research aim, we will investigate the function of attention getters in gesture sequences to clarify whether they are used to initiate a communicative interaction (anticipatory strategy) or as a reaction to a non-responding recipient (elaborative strategy).

2. Methods

2.1 Individuals

The present work uses observational data collected by Liebal et al. (2006). A total of 16 orangutans housed in two different captive groups (Zurich Zoo, Switzerland and Leipzig Zoo, Germany) was observed.

Table 1 provides an overview of the names, sex, and age classes of all individuals. The Leipzig group lived in an enclosure containing indoor and outdoor areas, which were enriched with several trees, ropes and platforms. Observations were conducted from the visitor areas and platforms above the indoor and outdoor enclosures. The Zurich group, housed in two connected indoor cages containing trees, robes and sleeping nests, was observed through glass windows from the visitor area. No changes in the daily routine of the apes were required.

Table 1. Individual characteristics and overview of kind and number of gesture sequences produced (Individuals from Zurich are presented in bold)

Individual	Age class	Sex	Number of sequences consisting of						Total of sequences	Total of single gestures
			two gestures	three gestures	four gestures	five gestures	six gestures	eight gestures		
Kila	infant	female	4	1	0	0	0	0	5	29
Xira	juvenile	female	5	3	1	1	0	1	11	81
Padana	juvenile	female	10	2	0	0	1	0	13	71
Sahli	subadult	female	9	0	1	0	0	0	10	89
Tuah	subadult	female	4	1	2	0	0	0	7	101
Toba	subadult	female	10	2	2	0	1	0	15	75
Pongo	adult	male	0	0	0	0	0	0	0	32
Lea	adult	female	1	1	0	0	0	0	2	48
Timor	adult	female	2	0	0	0	0	0	2	18
Selatan	adult	female	0	1	0	0	0	0	1	25
Oceh	adult	female	5	0	0	0	0	0	5	62
Djaro	adult	female	0	0	0	0	0	0	0	22
Bimbo	adult	male	0	0	0	0	0	0	0	12
Dunja	adult	female	4	2	0	0	0	0	6	127
Pini	adult	female	3	1	0	0	0	0	4	58
Walter	adult	male	5	0	0	0	0	0	5	61
Total			62	14	6	1	2	1	86	911

2.2 Observational methods

Each of the 16 individuals was observed for 10 hours using focal animal sampling (Altmann 1974) resulting in a video corpus of 160 hours observational data (Liebal et al. 2006). Each focal animal was selected randomly and observed for 15 minutes, before the next focal bout with a different individual started. If the focal animal moved outside the range of vision, the recording was stopped, and if it did not return within five minutes, the next session with a new individual was started. Daily observations took place between 08.30 am and 06.00 pm with the observation bouts equally distributed over the daily activity period of the apes.

2.3 Definitions

Gestures were defined as expressive, mechanically ineffective movements of limbs or head and body postures. For the definition and classification of gestures we used the gestural ethogram established by Liebal and colleagues (2006).

2.4 Coding procedure

In addition to the identities of the sender and recipient, we coded for each recorded gesture (1) its modality (tactile or visual), (2) the functional context in which it occurred, (3) the recipient's prior attentional state with respect to the sender, (4) whether and what kind of reaction the gesture elicited in the recipient, and (5) whether it occurred simultaneously with another gesture or was part of a gesture sequence. If one of these variables could not be coded, it was labeled as unknown (see Liebal et al. 2006).

Following the methodology of Liebal and colleagues (2004a), two gestures were part of a sequence if (1) both the sender and recipient were the same during this interaction (2) the functional context did not change, and (3) the production of the second gesture started within five seconds after the preceding gesture.

2.5 Reliability

Reliability was already established in the study by Liebal et al. (2006), where a second rater coded 20% of the data. Cohen's Kappa was used to measure their agreement in the types of gesture. Inter-rater agreement was $K = 0.77$, indicating a 'good' level of agreement (Altman 1991).

2.6 Statistics

Unless otherwise stated, the results are presented as mean proportions. Non-parametric statistics were used when degrees of freedom were below ten. In general, when comparing the use of single gestures and gesture sequences, only individuals who actually produced gesture sequences were included in the analyses.

3. Results

3.1 General overview

A total of 1128 gestures were recorded, including 740 (65.6%) tactile gestures and 388 (34.4%) visual gestures (see Table 2 for a listing of gestures observed). 217 of these gestures (19.2%) were part of one of 86 sequences. Table 1 presents a summary of the frequencies of single gestures and gesture sequences observed for each individual.

Non-adult individuals produced significantly more gesture sequences than adult individuals ($F(2,12) = 18.36$, $p < .001$) However, there was no difference between those age groups in their use of single gestures ($F(2,12) = 2.64$, $p = .11$).

Table 2. Listing and brief descriptions of the gestures observed

	Gesture	Code	Description
Tactile gestures	*Bite in hand*	BiH	Gently biting in recipient's hand
	Embrace	Em	Approaching and putting one or both arms around the body of the recipient
	Formal bite	FB	Touching the recipient with open mouth on any body part and biting it with low intensity
	Gentle touch	GT	Gently touching the recipient with hand or foot on any body part
	Lip touch	LT	Approaching the recipient with the face, touching the recipient's mouth with lips
	Nudge	Nu	Touching the recipient with single fingers, knuckles or fist
	Pull	Pl	Grasping any body part of the recipient by hand or foot and then performing a short, forceful movement with it
	Push	Pl	Pushing away any body part of the recipient with a short, vigorous movement
	Put face on	PFoF	Approaching the face of the recipient and taking the recipient's mouth into a slightly open mouth
	Put hand on	PHoH	Putting flat hand on the head or back of the recipient and leaving it there
	Slap	Sl	Hitting the recipient with a flat hand at any body part
	Throw object	TO	Throwing an object at another individual
	Touch with genital region	TwGR	Hanging in front of other individual and touching its face with genital region

(*Continued*)

Table 2. (Continued)

	Gesture	Code	Description
Visual gestures	*Approach face*	AF	Approaching the face of another individual staring at its mouth; sometimes eye contact
	Bite intention	BI	Indicating a rapid movement with the upper part of the body and the 'open mouth' towards another individual, but stopping without physical contact
	Extend arm	EA	Extending arm towards another individual; palm of the hand is not directed upwards
	Hand shake	HS	Shaking hands around the wrist
	Headstand	Hs	Turning vertically and standing on head and shoulders
	Hit intention	HI	Performing a movement with the arm as if to hit another individual; without physical contact
	Hold hand in front of the mouth	HHifotM	Putting the extended arm with the palm directed upwards in front of the mouth of another individual
	Jerking body movements	JBM	Hanging in front of another individual, moving the body up and down by angling the arms repeatedly.
	Offer arm with food pieces	OA	Chewing food into the fur of the arm and then extending the arm to present the food
	Offer body part	OBP	Offering any body part for grooming
	Offer food	OF	Extending arm with a piece of food in the hand, holding it in front of the mouth of another individual
	Present genitals	PG	Sitting or hanging in front of another individual, raising the posterior to present the genital region
	Present object	PO	Presenting an object by extending the arm with the object in the hand/s or by hiding under an object
	Shake object	SO	Shaking an object
	Wave arm	WA	Extending the arm, waveing it horizontally in front of the own body

Figure 1 shows the proportions of sequences as a function of the number and types of gestures combined and demonstrates that the frequency of gesture sequences decreased with the increasing number of gestures per sequence. The majority of gesture sequences consisted of only two gestures (72.1%, $N = 62$). Overall, about two thirds of the 86 sequences were repetitions of the same gesture (N = 57, 66.3%).

3.2 Functional contexts

Figure 2 shows the mean proportions of gesture sequences and single gesture in each functional context. Most gesture sequences were produced in the contexts

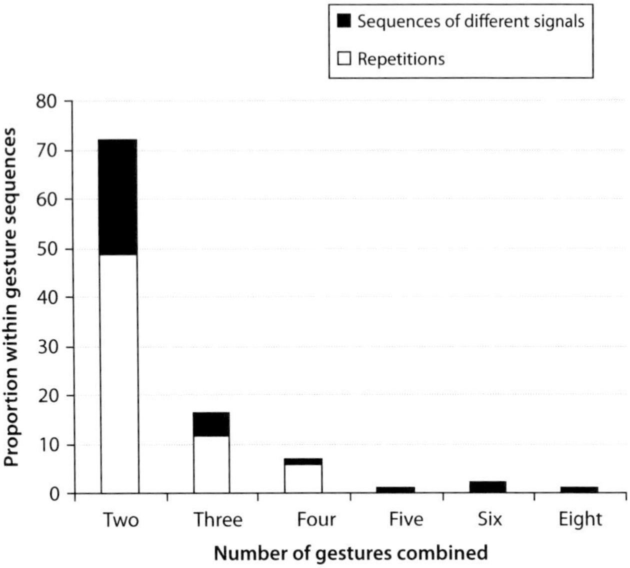

Figure 1. Proportion of gesture sequences as a function of number and type of gestures combined

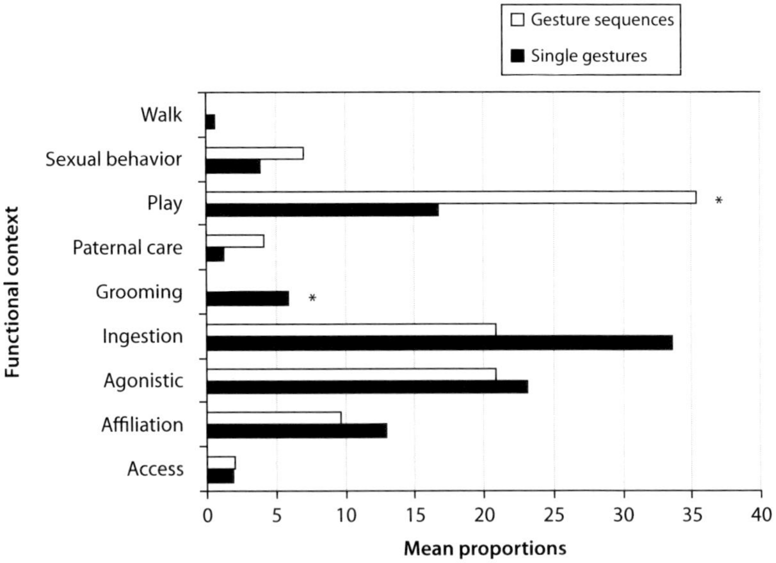

Figure 2. Mean proportions of gesture sequences compared to single gesture within each functional context (stars mark a significant difference between single gestures and gesture sequences regarding the mean proportion in the respective context)

of play (35.40% ± 35.97), ingestion (20.89% ± 24.29) and the agonistic context (20.85% ± 30.35). Compared to single gestures, gesture sequences were used more often in play (paired-samples t-test: $t(11) = -0.23$, $p < .05$), but less often in the context of grooming, in which the orangutans were never observed using gesture sequences (paired-samples t-test: $t(11) = 0.28$, $p < .05$). In all other contexts, the proportional use of single gestures compared to gesture sequences did not differ significantly (paired sample t-test: $df = 11$, all p-values > .05).

3.3 The use of gesture modalities as a function of the recipients' attentional state

Tactile gestures are efficient regardless of whether the recipient is attending or not, whereas visual gestures are only appropriate when the recipient is attentive. We analyzed whether senders considered the recipients' attentional states when producing the first gesture of a sequence. Overall, orangutans directed their first gesture significantly more often to attending than to not attending recipients (paired sample t-test, $t(12) = 4.04$, $p < .01$). Concerning gesture modality, tactile gestures were used significantly more frequent as first gestures of a sequence than visual gestures (paired sample t-test: $t(12) = -3.95$, $p < .01$). In addition, tactile gestures were used exclusively to non-attending recipients (100%), but also towards attending recipients (77.5% ± 29.3). Thus, the percentages of tactile gestures did not vary significantly depending on whether the recipient was attending or not (paired sample t-test, $t(12) = 0.12$, $p = .911$). In addition, only 22.5% (± 29.3) of gestures produced towards an attentive recipient were visual. However, contrary to tactile gestures they were exclusively directed towards an attending recipient, resulting in a significant difference between gesture modality with visual gestures being more often directed towards attentive than towards non-attentive recipients (paired sample t-test, $t(12) = 2.61$, $p < .05$).

3.4 Persistence

3.4.1 *Behavior of recipients*

We analyzed whether gesture sequences were more likely to emerge if there was no response to the first gesture of a sequence. We did not find a significant difference between the mean proportion of responses to single gestures (46.42% ± 13.53) compared to the first gestures produced in sequences (34.63% ± 28.69; paired-samples t-test: $t(12) = -1.43$, $p = .178$). We then looked at specific types of responses. First, we analyzed whether there were more responses that were more likely to terminate an interaction (e.g. the leaving of the recipient) after single gestures compared to initial gestures. There was no significant difference (paired-samples t-test: $t(11) = 0.42$, $p = .680$). Second, we compared whether a sender was more likely to stop gesturing if the recipient reacted with a context-specific behavior (such as play or grooming) indicating that the sender reached his/her communicative goal. Again, we did not find a significant difference between single and initial gestures (paired-sample t-test: $t(11) = 0.1530$, $p = .154$). Overall, these results indicate that there is no difference between the recipients' reaction to single gestures compared to initial gestures of a given sequence.

3.4.2 *Repetition and elaboration*

We analyzed in more detail whether the recipients' reaction to a first gesture influenced the senders' choice for a repetition of the same gesture or the switch to another one. This distinction is important, because when a gesture failed to achieve a communicative goal, the repetition of this gesture might reflect the sender's persistence. The modification of communicative means however, that is using a different gesture, would additionally indicate an elaborative use. Most crucial for the classification of an individual's communication as an act of persistence is the sender's consideration of whether his communicative goal has already been reached or not. When a communicative goal has already been achieved, the use of subsequent gestures for the same goal is unnecessary.

We found that when there was no response to the first gesture of a sequence, orangutans mostly repeated the same gesture (70.6% ± 25.8), switched to another gesture of the same modality in 15.3% (± 17.6), or to a gesture of a different modality in 14.0% (± 15.6) of the sequences. However, orangutans did not switch more often to a different gesture (same or different modality) when there was a response of the recipient compared to when there was no response (Wilcoxon test, $z = -0.985$, $p = .325$).

To summarize, our results showed that gesture sequences of orangutans did not emerge as a consequence of a lack of response to the initial gesture. The majority of gesture sequences were repetitions of the same gesture, regardless of the recipient's response. In addition, when the first gesture had failed to reach the communicative goal, the sender did not modify the gesture more often compared to when he had reached his goal. Therefore, gesture sequences of orangutans are not indicators of persistent gesture use to elicit the recipient's response.

3.4.3 *Attention getting gestures: Anticipatory and elaborated use*

Next, we analyzed whether orangutans used particular gestures either as first gesture of a sequence to attract the attention of the recipient, or alternatively to elaborate their communication (e.g. when there was no response of the recipient towards a previous gesture). We conducted three sets of analyses.

The first set is concerned with the question whether tactile gestures served as attention getters given that they can be perceived regardless of whether the recipient is attending or not. In the second set, we included attention getters, which were used as single gestures when the recipient was not attending. This selection was based on the assumption that those gestures might be especially effective to attract recipients' attention. In a third set, we identified more and less efficient gestures based on the response of recipients and investigated whether more efficient gestures were used in any strategic way.

3.4.3.1 *Gesture modality.* In terms of an anticipatory use of gestures, we focused on the first two gestures of a sequence. A tactile gesture was followed by another tactile gesture in more than three quarters of all sequences (77.7% ± 22.8), while a combination of two visual gestures occurred much less often (10.5% ± 16.9). It is important to take into account that this analysis included both repetitions of the same gesture as well as the switch to another gesture of the same modality. However, the crucial combination indicating that an attention getting gesture was used first to attract the recipient's attention followed by a visual gesture was observed in only 4 cases (6.5% ± 14.9). In only eight cases a tactile gesture actually resulted in attracting the attention of a previously inattentive recipient to the gesturing individual. Furthermore, in only two out of these eight cases, the subsequent gesture was a visual gesture. In addition, when there was no response to a first, visual gesture, the following gesture was not more often a tactile one (Wilcoxon test, $z = -1.414$, $p = .157$). Therefore, tactile gestures were not used to elaborate communication when a first visual gesture failed to reach the communicative goal.

3.4.3.2 *Gestures used for inattentive recipients.* We identified 13 gesture types, which were (as single gestures) directed towards an inattentive recipient. In sequences, these gestures were equally often used in the first position of sequences as other gestures (independent-samples t-test, $t(13.73) = 2.00$, $p = .065$) and they were not used more often either as first or second gesture of a sequence (paired-sample t-test, $t(12) = 0.00$, $p = 1$). Furthermore, they were not used more often if there was no response to the initial gesture of a sequence (paired-samples t-test, $t(12) = -0.36$, $p = .723$). These results suggest that these gestures were neither used in an anticipatory fashion as introductory attention getters, nor in terms of elaboration.

3.4.3.3 *Efficient gestures.* Another way to look at the strategic use of gesture sequences is to not only differentiating gestures in terms of their modality or their implementation as single gesture, but based on their efficacy. Thus, the purpose was to analyze whether orangutans would switch to a more efficient gesture in case there was no response of the recipient to the sender's first gesture. To identify efficient gestures, we calculated the average proportions of reactions to single gestures, including a change in the attentional state as well as specific responses for each gesture. We used the median split criterion to accordingly group gestures into less efficient (response in less than 60.5%) and more efficient gestures (response in more than 60.5%).

In terms of an anticipatory use, gestures classified as more or less effective were equally often used as first gesture in a sequence (independent-samples t-test: $t(19) = 0.82$, $p = .423$). More effective gestures were equally often used in the first and the second position of sequences (paired-samples t-test, $t(10) = 0.52$, $p = .617$).

Furthermore, in terms of an elaborated gesture use, there was no difference in the use of more efficient gestures in the second position of a sequence when there was no response compared to when there was a response to the first gesture (paired-samples t-test, $t(10) = -0.10$, $p = .923$). This result indicates that more efficient gestures were not strategically used to elaborate gesture use in case there was no response to the sender's first gesture.

Subsequent sequential analyses supported the findings that attention-getting gestures were neither used in an anticipatory way as first gesture of a sequence nor in an elaborated manner in case there was no response to the first gesture. Thus, the analysis of the first two gestures of a sequence showed that the transition probability that a more efficient gesture is followed by a less effective one was only 0.29. This result indicates that more efficient gestures were not used in an anticipatory way. On the other hand, when there was no reaction to a first, less efficient gesture, the transition probability to a second, more efficient gesture was still only 0.11. This result thus supports the previous finding that the sender switched to more efficient gestures in case there was no response to the first gesture. Furthermore, the transition probabilities did not differ significantly depending on whether there was a reaction to the first gesture or not (Breslow-Day test: $\chi = 0.17$, $p = .68$). Regardless of the recipient's behavior, the most frequent pattern of gesture combination was an efficient gesture followed by another efficient gesture and a not efficient gesture by a not efficient one. When there was no response to a first gesture, the transition probability for a more efficient gesture followed by another one was 0.73. For a less efficient gesture followed by another less efficient one, the transition probability was 0.89 (Pearson's Chi-square test f: $\chi = 9.53$, $p < .01$). When there was a response to the first gesture, the patterns were the same (transition probability: for more effective to more effective = 0.68, for less effective to less effective = 0.86; Pearson's Chi-square test: $\chi = 16.37$, $p < .001$). To summarize, our results show that more efficient gestures were not used strategically, neither in terms of elaboration nor anticipatorily to attract a communicative partner's attention.

4. Discussion

The aim of this paper was to investigate orangutans' use of gesture sequences in their interactions with conspecifics with a special emphasis on the question whether sequences are used in a strategic way to achieve a particular goal. We intended to explain under which circumstances gesture sequences emerge at all. To do so, we investigated whether orangutans combine gestures to attract the attention of others to a subsequent gesture and we examined their gesture use in terms of persistence in case there was no satisfying response of the recipient.

We found that approximately 20% of orangutans' gestures were part of a combination, which consisted of mostly two-gesture sequences involving the same gesture. Gesture sequences did not emerge because the recipient ignored the initial gesture, since the sender continued to gesture regardless of the recipient's response. This result indicates that orangutans' gesture sequences are not the result of persisting in communicating. Furthermore, there was no evidence supporting the hypothesis of a strategic gesture use to manipulate the recipient's behavior in an elaborated or anticipatory way.

Hobaiter and Byrne (2011) argue that the use of gesture sequences in chimpanzees can be explained by (1) persistence, (2) inter-individual adjustment, and (3) the development of the effective handling of the gestural repertoire in young individuals. As mentioned above, persistence could be ruled out as explanation for orangutans' use of gesture sequences. Inter-individual adjustment, which means the continuous regulation of tempo and nature of social interactions, is an interesting suggestion, but our data do not allow the testing of this hypothesis. However, Hobaiter and Byrne's (2011) main argument is that young individuals have to learn which gestures are effective for communicative purposes. Therefore, young chimpanzees use different types of gestures in series to assure the recipient's response. Gradually, they learn which gestures are effective and as a result, the need to use sequences decreases with increasing age (called *repertoire-tuning*; Hobaiter & Byrne 2011). Contrary, younger orangutans in the present study used more gesture sequences than adults, consisting mainly of repetitions of the same gesture. Therefore repertoire-tuning seems not a likely explanation for the use of sequences in orangutans.

As previously mentioned, both persistence and attention getting behavior are indicators for goal directed communication. What does the absence of these important characteristics tell us about orangutans' gestural communication? Compared to single gestures, orangutans used significantly more gesture sequences in the context of play, but significantly less in the context of grooming. Contrary to an episode of grooming, which is an even-tempered interaction, play involves high levels of arousal and activity. Therefore, the emergence of gesture sequences might not reflect any strategic use, but might simply be a result of the individual's arousal in particular contexts of gesture use. Therefore, the fact that senders often continued to gesture regardless of the recipient's behavior indicates that the production of gesture sequences might be the result of the sender's arousal, largely unaffected by the actual outcome of the (first) communicative attempt. Interesting in this context is that young individuals produced significantly more gesture sequences (but not single gestures) than adults. Furthermore, as opposed to all females, only one of the four adult males produced gesture sequences. This fits nicely with orangutans' social organization in the wild (Delgado & van Schaik; 2000). Long-lasting relationships are rare and apart from mothers with their offspring, sexual consort-ship, and groups of immature individuals, orangutans always maintain a pronounced inter-individual distance (Rijksen & Meijaard 1999).

Consequently, both the prerequisites and the opportunity to communicate with others are very limited, especially in form of gestural communication that requires close proximity between communication partners.

Thus, although orangutans can potentially live in bigger groups both in zoos and in the wild, the use of gesture sequences in captive individuals might still reflect their (semi-)solitary nature and the relatively low motivation to interact with others.

As opposed to gesture use between conspecifics, Cartmill and Byrne (2007) found that orangutans in interactions with humans persist if the human does not react at all and elaborate their gesture use if the human does not respond in the "expected" way. To apply our alternative explanation and to argue that the orangutans were aroused differently in the three conditions – no response, inappropriate response, appropriate response – resulting in different patterns of gesture use seems not to be compelling. However, the different findings of these two studies refer to two issues that are important in the area of gesture research on non-human primates, which are often not considered to the necessary extent. First of all, gestures produced by apes in interactions with humans are of a very specific nature: they are often limited to the food context, with the ape requesting a piece of food from a human. In these scenarios, orangutans and other great apes often produce gestures they would not use when interacting with conspecifics given the special setup of the experimental situation with the ape being spatial separated from the human. As a consequence, apes produce *pointing* and *begging* gestures, (which are not frequently used when interacting with conspecifics) to use the human as social tool to obtain the reward (Liebal & Call in press).

Second, contrary to experimental studies (e.g. Cartmill & Byrne 2007), in observational studies it is not always easy to specify the desired communicative goal of an individual in a reliable way. In interactions with conspecifics, it is much more difficult to measure whether the goal of the gesturing individual has actually been met by the response of the recipient.

Although there is evidence that orangutans are persistent in their communicative attempts in interactions with humans, we found no evidence for persistence in their gestural interactions with conspecifics. This suggests that the motivation to communicate with conspecifics might be very different from the motivation (and need) to communicate with humans, since the interactions with humans mostly involve food rewards. Whether this motivational aspect affects exclusively the gestural communication or is a general phenomenon can not be resolved by the current findings, but has to be the subject of further research.

In summary, our study suggests that opposed to chimpanzees, orangutans' use of gesture sequences in interactions with conspecifics does not result from the sender's persistence to achieve a communicative goal. Further, we found no evidence for the strategic use of specific gestures in terms of elaboration, as found for chimpanzees' and orangutans' interactions with humans or in terms of attention getting behavior (Cartmill & Byrne 2007; Leavens et al. 2005).

We thus suggest that gesture sequences of orangutans' do not represent a strategy to get another individual to react, but mainly emerge as a result of an individual's high level of arousal. The difference between chimpanzees and orangutans in their use of gesture sequences most likely has its roots in their quite different social systems. However, the different findings resulting from studies of interactions between conspecifics on the one hand and human – ape interactions on the other hand suggest that both observational and experimental studies are essential to comprehensively describe the communicative patterns and skills of non-human primates.

Acknowledgements

Many thanks to the keepers and administrative staff of the Zoo Zurich, in particular Paul Bosshardt, René Erni, and Robert Zingg as well as the team of Frank Schellhardt at Leipzig Zoo. We also want to thank Daniel Stahl for statistical advice, and Daniel Haun, Trix Cacchione, and Richard Moore for comments on earlier versions of this manuscript. This project was funded by the Max Planck Society and the Volkswagen – Stiftung, Germany.

References

Altman, Douglas G. 1991. *Practical statistics for medical research*. Chapman and Hall: CRC Press.

Altmann, Jeanne. 1974. "Observational study of behavior: Sampling methods." *Behaviour* 49: 227–267.

Bates, Elisabeth, Benigni, Laura, Bretherton, Inge, Camaioni, Luigia and Volterra, Virginia. 1979. *The emergence of symbols: Cognition and communication in infancy*. New York: Academic Press.

Bates, Elisabeth, Camaioni, Luigia and Volterra, Virginia. 1975. "The acquisition of performatives prior to speech." *Merrill-Palmer Quarterly* 21:205–224.

Bruner, Jeromy. 1981. "Intention in the structure of action and interaction." In *Advances in Infancy Research* (Vol. 1), L. Lipsitt (ed.), 41–56. New Jersey: Ablex, Norwood.

Call, Josep and Tomasello, Michael. 2007. *The gestural communication of apes and monkeys.* Mahwah, NJ: Lawrence Erlbaum.

Cartmill, Erica A. and Byrne, Richard W. 2007. "Orangutans modify their gestural signalling according to their audience's comprehension." *Current Biology* 17:1345–1348.

Cartmill, Erica A. and Byrne, Richard W. 2010. "Semantics of primate gestures: intentional meanings of orangutan gestures." *Animal Cognition* 13:793–804.

Corballis, Michael C. 2002. *From hand to mouth: The origins of language.* Princeton: Princeton University Press.

Delgado, Roberto A., Jr. and van Schaik, Carel. 2000. "The behavioral ecology of the orangutan (*Pongo pygmaeus*): A tale of two islands." *Evolutionary Anthropology* 9:201–218.

Genty, Emilie, Breuer, Thomas, Hobaiter, Catherine and Byrne, Richard W. 2009. "Gestural communication of the gorilla (*Gorilla gorilla*): Repertoire, intentionality and possible origins." *Animal Cognition* 12 (3):527–546.

Genty, Emilie and Byrne, Richard W. 2010. "Why do gorillas make sequences of gestures?" *Animal Cognition* 13 (2):287–301.

Hobaiter, Cathereine and Byrne, Richard W. 2011. "Serial gesturing by wild chimpanzees: its nature and function for communication." *Animal Cognition* 14 (6):827–838.

Hostetter, Autumn. B., Cantero, Monica and Hopkins, William D. 2001. "Differential use of vocal and gestural communication by chimpanzees (*Pan troglodytes*) in response to the attentional state of a human (*Homo sapiens*)." *Journal of Comparative Psychology* 115:337–343.

Kendon, Adam. 2004. *Gesture: Visible action as utterance.* Cambridge: Cambridge University Press.

King, Barbara J. 2004. *The dynamic dance. Nonvocal communication in African great apes.* Cambridge, Massachusetts, London: Harvard University Press.

Leavens, David A., Russell, Jamie L. and Hopkins, William D. 2005. "Intentionality as measured in the persistence and elaboration of communication by chimpanzees (*Pan troglodytes*)." *Child Development* 76:291–306.

Liebal, Katja and Call, Josep *in press*. "The origins of nonhuman primates' manual gestures." *Philosophical Transactions of the Royal Society B.*

Liebal, Katja, Call, Josep and Tomasello, Michael. 2004a. "Use of gesture sequences in chimpanzees." *American Journal of Primatology* 64 (4):377–396.

Liebal, Katja, Pika, Simone, Call, Josep and Tomasello, Michael. 2004b. "To move or not to move: How apes alter the attentional states of humans when begging for food." *Interaction Studies* 5:199–219.

Liebal, Katja, Pika, Simone and Tomasello, Michael. 2006. "Gestural communication of orangutans (*Pongo pygmaeus*)." *Gesture* 6:1–38.

Maestripieri, Dario. 1999. "Primate social organization, gestural repertoire size, and communication dynamics: a comparative study of macaques." In *The origins of language. What nonhuman primates can tell us,* Barbara J. King (ed.) 55–77. Santa Fe, NM: School of American Research Press.

Pika, Simone and Liebal, Katja. 2012. "Introduction: Developments in primate gesture research." In *Developments in Primate Gesture Research*, Simone Pika and Katja Liebal (eds) 1–11. Amsterdam/Philadelphia: John Benjamins Publishing Company.

Pika, Simone, Liebal, Katja and Tomasello, Michael. 2003. "Gestural communication in young gorillas (*Gorilla gorilla*): Gestural repertoire, learning, and use." *American Journal of Primatology* 60: 95–111.

Pika, Simone, Liebal, Katja and Tomasello, Michael. 2005. "Gestural communication in subadult bonobos (*Pan paniscus*): Gestural repertoire and use." *American Journal of Primatology* 65: 39–61.

Rijksen, H.D. and Meijaard, Erik, E. 1999. *Our vanishing relative: The status of wild orangutans at the close of the twentieth century*. Dordrecht, NL: Kluwer.

Smuts, Barbara B., Cheney, Dorothy. L., Seyfarth, Robert M., Wrangham, Richard W. and Struhsacker, Thomas T. 1987. *Primate Societies*. Chicago: University of Chicago Press.

Tanner, Joanne E. 2004. "Gestural phrases and gestural exchanges by a pair of zoo-living lowland gorillas." *Gesture* 4: 1–24.

Tempelmann, Sebastian, Kaminski, Juliane and Liebal, Katja. 2011. "Focus on the essential: All great apes know when others are being attentive." *Animal Cognition* 14 (3): 433–439.

Tomasello, Michael, Call, Josep, Nagell, Katherine, Olguin, Raquel and Carpenter, Malinda. 1994. "The learning and the use of gestural signals by young chimpanzees: A transgenerational study." *Primates* 35: 137–154.

Tomasello, Michael, Call, Josep, Warren, Jennifer, Frost, Thomas, Carpenter, Malinda and Nagell, Katherine. 1997. "The ontogeny of chimpanzee gestural signals: A comparison across groups and generations." *Evolution of Communication* 1: 223–253.

van Schaik, Carel. P. 1999. "The socioecology of fission-fusion sociality in orangutans." *Primates* 40 (1): 69–86.

Handedness for manual gestures in great apes

A meta-analysis

William D. Hopkins,[1,2] Simone Pika,[3] Katja Liebal,[4]
Amanda Bania,[5] Adrien Meguerditchian,[2] Molly Gardner &
Steven J. Schapiro

[1]Department of Psychology, Agnes Scott College, Decatur, Georgia, United States / [2]Division of Developmental and Cognitive Neuroscience, Yerkes National Primate Research Center, Atlanta, Georgia, United States / [3]Humboldt Research Group "Comparative Gestural Signalling", Max Planck Institute for Ornithology, Eberhard-Gwinner-Strasse, 82319, Seewiesen, Germany / [4]Department of Psychology, Freie Universitat Berlin, Habelschwerdter Allee 45, 14195 Berlin, Germany / [5]The National Zoo, Washington, D.C. Department of Veterinary Sciences, The University of Texas M.D. Anderson Cancer Center, Bastrop, Texas

Left hemisphere dominance for language and speech is a prominent feature of the human brain. Some have suggested that left hemisphere lateralization for language in modern humans evolved from an existing lateralized system for gestural communication in the common ancestor of humans and apes. Here, we conducted a meta-analysis of the data on handedness for manual gestures in great apes to test whether species-level asymmetries in hand use were evident and if these asymmetries were consistent across species. Hand preference data for manual gestures were summarized for published and unpublished data in the literature. For observational data, we found that bonobos, chimpanzees, gorillas and orangutans all showed group-level right handedness for manual gestures. For experimental data, group-level right handedness was only found for chimpanzees, though sample sizes were relatively small in the remaining ape species. When hand preference data were combined across ape species, we found that right handedness was much more prevalent for auditory gestures, such as CLAPPING, compared to visual and tactile gestures. The results support the view that asymmetries in hand use for gestural communication was prevalent in the common ancestor of chimpanzees and humans and was possibly evident even earlier in primate evolution. Additional data are needed, particularly from wild apes to test for the presence of these asymmetries in subjects raised in their natural environment.

One of the most distinctive features of the human brain and behavior is hemispheric specialization (Annett 2002; Bradshaw & Rogers 1993; Hatta 2006; Toga & Thompson 2003). Two of the most robust manifestations of hemispheric specialization in humans are handedness and language. Specifically, though there is some cultural variation, a significant majority of humans report preferring to use the right hand for a variety of motor actions (Annett 2002; Porac & Coren 1981), which suggest that the left hemisphere is dominant for motor actions. In addition, clinical and experimental studies, including recent functional imaging techniques, have consistently shown that the left inferior frontal gyrus and posterior temporal cortex are involved in the production and perception of both spoken and signed languages (Binder et al. 1997; Hickok, Bellugi & Klima 1998), as well as spontaneous manual gestures (Horwitz 2003; Kimura 1993).

There is also a relationship between handedness and language lateralization. Early studies using the sodium-amytal (Wada) test demonstrated that approximately 96% of right-handed individuals were left hemisphere dominant for language in contrast to only 70% of the left-handed individuals (Rasmussen & Milner 1977). During the Wada test, one hemisphere is temporarily anesthesitized while patients are awake and talking. Speech is typically arrested when the left but not right hemisphere is anesthestized, particularly among right-handed individuals. More recent studies using transcranial Doppler blood flow sonography have similarly shown that about 96% of strongly right-handed individuals show left hemisphere dominance for language contrasted with about 73% for left-handed persons (Knecht, Deppe et al. 2000; Knecht, Drager et al. 2000). Thus, though the majority of left- and right-handed individuals show left hemisphere dominance for language, the proportion is significantly higher among right-handers.

The prevalence of right handedness and left hemisphere dominance for language, as well as their statistical association, has led some to suggest that these two behavioral asymmetries evolved together after the split between humans and chimpanzees (Cashmore, Uomini & Chapelain 2008; Crow 2004; Warren 1980). The data principally used to support this evolutionary interpretation has been the historical lack of evidence of population-level asymmetries in hand preferences in non-human primates, and indeed, non-humans altogether (Cashmore et al. 2008; Crow 2004; Warren 1980). Related to these data were the general view that language and speech were unique to humans and were not present in other animals. Thus, theoretical discussions of the evolutionary origins of language often referenced the lack of evidence of population-level handedness in non-human animals as evidence in support of the uniqueness argument (Corballis 1992).

Recent studies in non-human animals, including primates, have begun to challenge the assumption that hemispheric specialization is unique to humans. There is now a large body of data suggesting population-level asymmetries for perceptual, motor and cognitive functions in a variety of distantly related vertebrates, such as birds, frogs, fish, just to name a few (Rogers & Andrew 2002; Vallortigara & Rogers 2005). For example, toads prefer to use their right paw to remove tape attached to their heads (Bisazza, Cantalupo, Robins, Rogers & Vallortigara 1996) and a number of bird species prefer to use one eye over another when discriminating between small pebbles and grains of food (Andrew 1991; Güntürkün 1997).

With specific reference to primates, though still somewhat debated, there is some evidence of population-level handedness, particularly among great apes. Population-level handedness refers to circumstances in which a statistically significant majority of individuals within a sample show the same hand preference. For instance, captive chimpanzees show population-level right handedness for throwing, manual gestures, coordinated bimanual actions such as bimanual feeding, leading limb in locomotion, and, to a lesser extent, simple reaching (Hopkins 2006; Hopkins, Russell, Lambeth & Schapiro 2007). In some of the more well studied wild chimpanzee communities, population-level left handedness has been found for termite fishing at Gombe (Lonsdorf & Hopkins 2005) while right-handedness has been found for leaf sponging (Bossou and Taï), ant dipping or fishing (Mahale and Bossou), simple reaching (Bossou) and bimanual feeding (Mahale) (Biro et al. 2003; Biro, Sousa & Matsuzawa 2006; Boesch 1991; Corp & Byrne 2004; Humle & Matsuzawa 2009; Marchant & McGrew 2007). Why there is group-level variation in handedness across different tool use tasks in wild chimpanzees is not clear but similar between-task variation is found in captive chimpanzees; therefore, it does not seem attributable to the rearing experiences of the apes. There are fewer reports of handedness in other captive and wild apes but some have documented evidence of population-level right handedness for bimanual feeding in wild mountain gorillas at Virunga and captive zoo-living lowland gorillas (Byrne & Byrne 1991; Meguerditchian, Calcutt, Lonsdorf, Ross & Hopkins 2010). Experimental studies of hand preferences for coordinated bimanual actions in captive apes have reported population-level right handedness in gorillas and bonobos (to a lesser extent) whereas population level left handedness was found in orangutans (Hopkins et al. 2011; Hopkins, Stoinski, Lukas, Ross & Wesley 2003; Rogers & Kaplan 1995). There is one report of population-level right handedness in the leading limb during brachiation in wild Bornean orangutans (Peters & Rogers 2007).

There is no doubt that the majority of studies on handedness in captive and wild apes have focused largely on manual actions that are non-communicative in function, yet theoretically many have attempted to link the evolution of hemispheric specialization with the emergence of language and speech (Bradshaw 1997; Bradshaw & Rogers 1993; Corballis 2002). Thus, if scientists are seeking to understand the evolution of hemispheric specialization in relation to language and speech, then there should be a greater focus on the study of lateralization in behaviors that are inherently communicative in function.

In this paper, we examine the evidence of handedness for manual gestural communication. Our theoretical interest in this topic stems from some evolutionary theories postulating that human language and speech evolved from an extant gestural communication system (Arbib 2005; Hewes 1976; Rizzolatti & Arbib 1998). With specific reference to hemispheric specialization and handedness, Corballis (2002) has suggested that early hominids had a left hemisphere specialization in manual gestures, that was subsequently co-opted in modern human evolution to support the emergence of speech and language. The fundamental question for the Corballis hypothesis is how far back the origin of left hemisphere specialization for gestures can be extended within Hominins. A number of studies in captive and wild apes have described complex and flexible gestural communication systems in all the great apes (Call & Tomasello 2007; Cartmill & Byrne 2007; Cartmill & Byrne 2010; Genty, Breuer, Hobaiter & Byrne 2009; Kalan & Rainey 2009; Leavens, Hostetter, Wesley & Hopkins 2004; Leavens, Russell & Hopkins 2005; Liebal, Call & Tomasello 2004; Liebal, Müller & Pika 2005; Liebal, Pika, Call & Tomasello 2004; Pollick & de Waal 2007) and therefore these are ideal model species' for potentially assessing the phylogeny of hemispheric specialization in relation to gestural communication. Here, we review existing findings and incorporate new unpublished data on gestural communication in great apes with the goal of (a) evaluating whether great apes show population-level handedness in manual gestures and (b) whether experimental or observational methods influence the expression of handedness in great apes.

1. Methods

1.1 Subjects

In total, handedness for manual gesture data were available in 498 individual captive apes including 80 bonobos (*Pan paniscus*), 336 chimpanzees (*Pan troglodytes*), 78 gorillas (*Gorilla gorilla gorilla*) and 30 orangutans (*Pongo pygmaeus*), respectively. The apes were living in either North American or European zoos or research facilities within the United States. Subjects ranged from two to 50 years of age.

1.2 Selection of studies

For the meta-analysis, individual data were computed from either published studies on handedness for multiple measures which included a category of gestural communication (Fletcher & Weghorst 2005; Harrison & Nystrom 2008; Hopkins et al. 2005; Meguerditchian, Vauclair & Hopkins 2010; Shafer 1993; Shafer 1997) or were unpublished data provided by the authors of this paper based on previous observational or experimental studies they have conducted on gestural communication in great apes (Liebal et al. 2004; Liebal, Pika & Tomasello 2006; Pika, Liebal & Tomasello 2003; Pika, Liebal & Tomasello 2005; Poss, Kuhar, Stoinski & Hopkins 2006). Shown in Table 1 are the sources of data, the species, the samples sizes and the method used to collect the hand preference data. The studies were divided as using either an experimental or observational method. For the experimental method, these studies involved the apes gesturing to request foods from a human experimenter who had placed the food items outside of the ape's home cage. For the most part, discrete trials were presented to each subject and the sum of the left and right hand responses were computed across trials. Thus, in these studies, the human experimenters explicitly sought to elicit hand use in the apes and a significant amount of effort was made to position the food in the central view of the ape so as not to bias hand use.

Table 1. Published and unpublished data analyzed in this study

Authors	Species	N	Method
Fletcher (2006)	Chimpanzee	16	Observational
Harrison and Nystrom (2008)	Bonobo	21	Observational
Harrison and Nystrom (2010)	Gorilla	7	Observational
Hopkins & de Waal (1985)	Bonobo	21	Observational
Hopkins (unpublished)	Bonobo	28	Experimental
Hopkins et al. (2005)	Chimpanzee	237	Experimental
Hopkins (unpublished)	Chimpanzee	145	Observational
Hopkins (unpublished)	Gorilla	9	Experimental
Hopkins (unpublished)	Orangutan	6	Experimental
Liebal et al. (2004)	Chimpanzee	19	Observational
Liebal et al. (2006)	Orangutan	16	Observational
Meguerditchian et al. (2010)	Chimpanzee	88	Observational
Meguerditchian et al. (2010)	Gorilla	10	Observational
Pika et al. (2005)	Bonobo	7	Observational
Pika et al. (2003)	Gorilla	16	Observational
Poss et al. (2006)	Gorilla	7	Experimental
Poss et al. (2006)	Orangutan	8	Experimental
Shafer (1997)	Bonobo	10	Observational
Shafer (1993)	Gorilla	36	Observational

For the observational studies, hand use for a variety of communicative manual actions was coded in the apes and the ethograms employed by Pika et al. (2003, 2005) and Liebel et al. (2004, 2006) were the most exhaustive and consistent between species. In these studies, manual actions scored for hand use were generally divided into larger classes of signals based on their modality which included auditory, tactile and visual gestures. Auditory gestures are defined as signals that produced a sound without physical contact and included behaviors such as SLAP, CHEST BEAT, CLAP, SLAP GROUND, and STOMP. Tactile gestures were defined as signals that included physical contact with the recipient such as EMBRACE, TOUCH, PULL, PUSH, POKE, GRAB and PUNCH, to name a few. Lastly, visual gestures consisted of signals that were visually directed toward the recipient but did not involve physical contact. These included behaviors such as HAND SHAKE, HAND CHUCK, OBJECT SHAKE, REACH (but without contact), FOOD BEGGING, EXTENDED ARM, and THROWING. For each of these behaviors, no attempt was made to restrict data collection to circumstances in which subjects were centrally positioned with respect to the target of their communication but rather hand use was noted without regard to positional factors. Furthermore, because the studies were observational and not necessarily focused on obtaining laterality data, no attempt was made to collect a minimum number of responses for each ape. Sampling techniques also varied across studies but the majority used focal sampling with some combined use of analyses of videotaped sequences of gestural communication.

1.3 Data analysis

For each subject, we computed a handedness index following the formula [(HI = (#R – #L) / (#R + #L)] where #R and #L reflect the frequency in right and left hand use. Positive HI values reflect right hand biases while negative values reflected left hand bias. In addition, we calculated binomial *z*-scores for each subject based on the frequency of right and left hand use. Based on these *z*-scores, we then classified subjects as either strongly left-handed ($z < -1.95$), moderately left-handed ($-1.95 < z < -1.0$), ambiguously-handed $-1.0 < z < 1.0$), moderately right-handed ($z => 1$ but $z <= 1.95$) and strongly right-handed ($z > 1.95$). For all analyses, we only included subjects that had a minimum of six observations. In our analyses, we assessed hand preferences for the experimental and observational data separately. For the experimental data, hand use data for subjects with six or more observations were available from 285 apes including 27 bonobos, 231 chimpanzees, 15 gorillas and 12 orangutans (see Table 2). For the observational methods, data from apes with six or more observations were available in 324 apes including 53 bonobos, 210 chimpanzees, 45 gorillas and 16 orangutans (see Table 2).

Table 2. Handedness distribution for the different methods of data collection

	#SL	#ML	#AB	#MR	#SR	%L	%R	R:L Ratio
Observational								
Bonobo	5	3	26	9	10	15	36	2.38
Chimpanzee	22	21	64	32	71	20	49	2.48
Gorilla	2	3	16	9	15	10	50	5.00
Orangutan	0	2	6	6	2	13	50	4:00
Experimental								
Bonobo	7	2	7	2	9	33	41	1.22
Chimpanzee	26	19	47	19	120	19	60	3.08
Gorilla	4	2	1	0	8	40	53	1.33
Orangutan	4	1	3	1	3	41	33	0.80

#SL = significantly left-handed individuals, #ML = moderately left-handed individuals, #AB = ambiguously-handed individuals, #MR = moderately right-handed individuals, #SR = significantly right-handed individuals. %L and %R were calculated by dividing the number of (SL + ML) or (SR + MR) by the total number of individuals (#SL + #ML + #AB + #MR + #SR). The R:L Ratio was computed by dividing %R by %L.

2. Results

2.1 Experimental data

For the HI values, a one-way analysis of variance revealed a significant difference between species $F(3, 281) = 3.497$, $p < .02$. Post-hoc analyses indicated that chimpanzees had significantly higher HI scores than orangutans. No other comparisons in means differed significantly between ape species. One sample *t*-tests within each species revealed population-level right handedness for chimpanzees $t(230) = 8.38$, $p < .001$ while none of the other great apes showed significant population-level biases (see Figure 1). The distribution of handedness for each species is shown in Table 2. To increase our statistical power, we combined the SL and ML individuals into one group labeled left-handed and the MR and SR individuals into a single group called right-handed. We then compared the four species across the three handedness categories (left-handed, ambiguously-handed and right-handed) using chi-square statistics. No species by handedness group differences were found. However, a chi-square goodness of fit test showed that handedness differed significantly from a random distribution $X^2(4, N = 285) = 165.96, p < .001$. Overall, there were more ambiguously-handed apes than all other preference patterns. However, among the strongly-lateralized $X^2(1, N = 181) = 53.06, p < .001$ apes, there were significantly more right- then left-handed individuals. In short, though many apes failed to show a bias in hand use, if they did show a significant preference, it was more prevalent for the right compared to left hand.

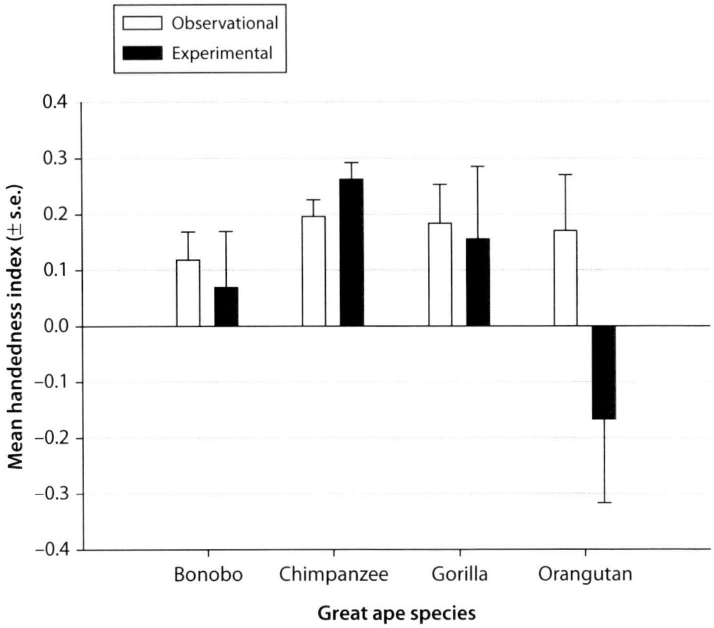

Figure 1. Mean HI scores (± s.e.) for all gestures observed in the observational and experimental conditions for each species

2.2 Observational data

For the HI values, a one-way analysis of variance failed to reveal significant differences between species $F(3, 320) = 0.496$, n.s. One sample t-tests within each species revealed population-level right handedness for bonobos $t(52) = 2.46$, $p < .02$, chimpanzees $t(209) = 4.88$, $p < .001$, gorillas $t(44) = 4.16$, $p < .001$ and orangutans $t(15) = 3.30$, $p < .005$ (see Figure 1). The distribution of handedness based on their z-scores for each species is shown in Table 2. To increase our statistical power, we combined the SL and ML individuals into one group labeled left-handed and the MR and SR individuals into a single group called right-handed. We then compared the four species across the three handedness categories (left-handed, ambiguously-handed and right-handed) using chi-square. As was the case for the HI measures, no species by handedness group differences were found. However, a chi-square goodness of fit test showed that handedness differed significantly from a random distribution $X^2(4, N = 324) = 90.41$, $p < .001$. Overall, there were more ambiguously-handed apes than all other preference patterns. However, among the moderately $X^2(1, N = 89) = 5.44$, $p < .02$ and strongly-lateralized $X^2(1, N = 123) = 39.84$, $p < .001$ apes, there were significantly more right- then left-handed individuals. In short, though many apes failed to show a bias in hand use, if they did show a moderate or significant preference, it was more prevalent for the right hand.

2.3 Repeat observations

Data were available in 138 chimpanzees and six bonobos for both observational and experimental methods. For this analysis, we sought to examine consistency on hand preference between the two methods when obtained in the same subjects. A paired sample t-test between HI scores in the experimental (Mean = .24) and observational (Mean = .22) conditions did not differ significantly. Moreover, one sample *t*-test on the HI scores revealed significantly population-level right handedness for both methods. Finally, we also found that the correlation between HI measures between the two methods were positive and significant, though admittedly weak ($r = .213$, df = 142, $p < .01$).

2.4 Gesture specific handedness

Though the types and frequency in hand use for different types of gestures varied substantially across the studies, in this analysis, we sought to assess whether hand preferences varied in relation to the type of gesture, as defined in very broad terms. As noted previously, several studies have made distinctions between auditory, visual and tactile gestures. Thus, for this set of analyses, within each study, we summed the number of left and right hand gestures across the different behaviors that were categorized as auditory, tactile or visual signals (see above). Limiting the analysis to subjects with six or more observations, significant population-level right handedness was found for auditory $t(104) = 5.21$, $p < .001$, visual $t(105) = 2.76$, $p < .01$, and tactile $t(105) = 3.18$, $p < .01$ gestures. The mean HI scores for each class of gestures is shown in Figure 2. Though right hand biases were evident for all three classes of signals, it is clear that lateralized hand use for auditory gestures were much more pronounced than for visual and tactile gestures. We also compared the HI scores between the four ape species for each category of gesture and found no significant differences.

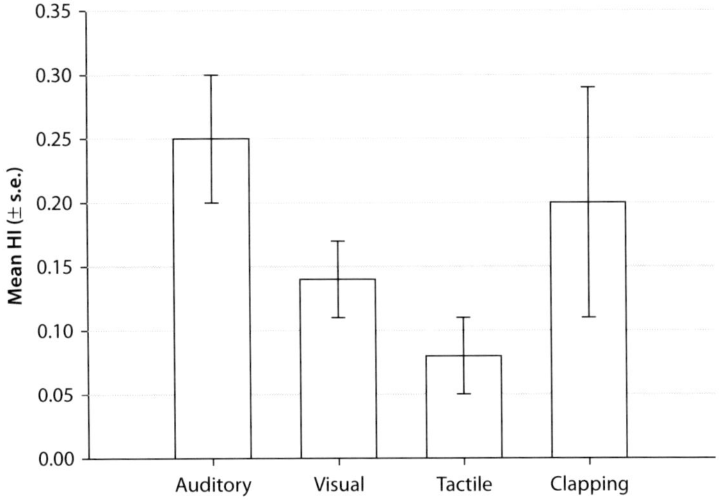

Figure 2. Mean HI scores (± s.e.) for the different categories of gestures measured in the observational studies.

Finally, one manual auditory gesture that was nearly exclusively recorded in chimpanzees was CLAPPING (there were data on 6 gorillas, Meguerditchian, unpublished), which was highly individual specific and nearly always directed toward humans in both laboratory and zoo settings (Fletcher 2006; Meguerditchian et al. 2010). We pulled this behavior out for separate analysis because it was used by many apes and has been shown to be used as an attention-getting behavior in captive apes (Leavens et al. 2004). Moreover, Fletcher (2006) reported that hand use for CLAPPING was strongly lateralized and therefore a potentially good measure of hemispheric specialization. The preferred hand for CLAPPING was defined as the hand that was positioned on top when the two hands come together during a CLAPPING bout and hand use data were available in 105 individuals. Interestingly, 87% of the sample showed at least 90% consistent hand use or, in other words, exclusive hand use for CLAPPING and, like other gestures, a significant population-level right hand bias was found $t(104) = 2.20$, $p < .03$ (see Figure 2).

3. Discussion

The aim of the present chapter was to summarize the available published and unpublished data on hand preference for manual gestures in great apes. The results of these analyses offer several insights into lateralization for gestural communication in great apes. First, independent of whether the data were collected using observational or experimental methods, great apes showed significant population-level right handedness but these results were most pronounced in chimpanzees and gorillas. Second, when considering the findings for observational compared to experimental methods, the results varied depending on the species. For observational methods, population-level biases were evident for all the great apes species and no significant differences were found between species. In contrast, for the structured experimental tasks, only chimpanzees showed a population-level asymmetry in hand preference and some significant differences were found between species. Lastly, when hand use for gestures were recorded the same subjects in both the observational and experimental context, hand preferences were weakly consistent and did not differ between settings.

When the results of this analysis are considered collectively, there are several important issues and points that arise and warrant some discussion. First, the fact that the apes were more consistently right-handed when based on the observational compared to the experimental data is somewhat surprising and, at the same time, very revealing. Specifically, intuitively, one might predict that asymmetries would be more prevalent under more controlled experimental circumstances where situational factors, such as position, can be better controlled or eliminated. In the case of this study, one would assume that during observational data collection, the hand used to gesture by the apes might be influenced by the position of the recipient of the gesture. This would presumably be randomly distributed across time and thereby induce a random distribution of handedness. By contrast, the degree of preferential use of the hands and expression of asymmetries was more pronounced in the observational compared to experimental conditions and was evident for several different types of gestures (see Table 2 and Figure 1). This suggests that future studies of laterality in intraspecies natural communication may reveal even more pronounced expression of handedness than non-communicative actions, particularly if positional factors are taken into consideration during the recording of hand use. Furthermore, the fact that hand preferences were more pronounced in natural communicative interactions of captive great apes suggests that measuring handedness for similar behaviors in wild populations of apes would potentially be a very fruitful area of investigation. This is particularly significant because some studies suggest substantial overlap in the gestural repertoires of wild and captive apes (Genty et al. 2009) and therefore this would strengthen the comparison of results between captive and wild apes, an approach rarely employed in existing studies of handedness for other types of manual actions.

The evidence of population-level handedness for manual gestures leads to the question of whether lateralization for communication is a more sensitive measure of hemispheric specialization than other manual actions that are not communicative in function. This is a difficult question to address in our sample of data because multiple measures of hand use were not available in all the subjects. Notwithstanding, available experimental data from captive chimpanzees, baboons and human children are consistent with the suggestion that manual gestures elicit stronger asymmetries than motor actions that are not communicative in function. Shown in Figure 3 are the mean HI scores for each species for manual gestures and simple reaching when collected in experimental setting. As can be seen, for all three species, the degree of right handedness is significantly higher for human children (Cochet & Vauclair 2010, n = 54), chimpanzees (Hopkins et al. 2005, n = 224), and baboons (Meguerditchian & Vauclair 2006, n = 190). It should also be noted that chimpanzees who preferentially gesture with the right hand have been found to have larger left hemisphere asymmetries in the inferior frontal gyrus (IFG) (Taglialatela, Cantalupo & Hopkins 2006) and planum temporale (PT) (Hopkins & Nir 2010), brain regions that correspond to the homologs of Broca's and Wernicke's areas in the human brain. In contrast, comparison of asymmetries in the IFG and PT do not differ between right- and left-handed apes when based on their preference for simple reaching or other non-communicative manual actions.

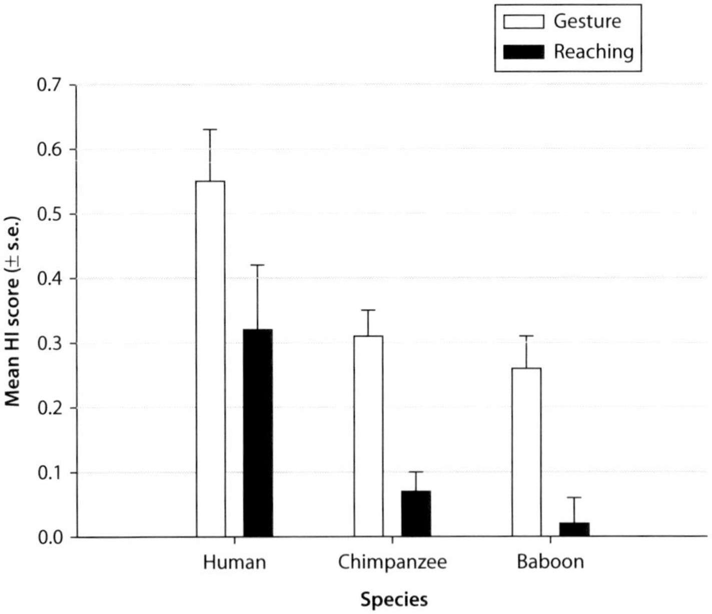

Figure 3. Mean HI scores (± s.e.) for manual gestures and simple reaching in humans, chimpanzees and baboons.

Though directional biases in hand preference were more prevalent for the right compared to left hand in the apes, it is important to recognize that many individuals were classified as ambiguously-handed or, in other words, they had no "significant" hand preference. Whether this reflects a biologically meaningful result or whether it is simply an artifact of the statistical and methodological approaches used in these studies is unclear. One issue that arises is what constitutes a "significant" hand preference. Many scientists studying handedness in non-human primates have adopted z-score cutpoints and critical values within the normal distribution to define subjects as being right-, left- or ambiguously-handed, not unlike the approach employed in this study (see Hopkins 1999; Hopkins & Cantalupo 2005; McGrew & Marchant 1997; Palmer 2002 for discussion). Using this approach, significant hand preferences are only evident for subjects with z-scores that exceed an alpha level of $p < .05$, which is correspondingly scores $<= -1.96$ or $=> 1.96$. By this criterion, only subjects at the extreme ends of the normal distribution are considered "significantly" left- or right-handed while all others are classified as ambiguously-handed. This is an exceedingly conservative criterion for determining handedness and, in point of fact, there are few if any studies on hand preferences in human adults or children that adopt such a rigid criterion. This simple fact makes comparisons in the distribution of handedness very difficult between human and non-human primate species. Further complicating the matter is the fact that z-scores are influenced by sample sizes, a particular limitation in this study. An example will make this point more apparent to the reader. Imagine you have hand preference data on two chimpanzees, one gestured three times with her left hand and seven times with her right while the second ape gestured 30 times with her left hand and 70 times with her right hand. The z-scores for these two apes would +1.26 and +4.00, respectively. Adopting the z-score criterion for classification of handedness, chimpanzee one would be considered ambiguously-handed while chimpanzee two would be classified as strongly right-handed, despite the fact that their handedness scores would be identical (HI = .40). Certainly, we would be more confident in the internal structure of handedness in chimpanzee two compared to one but, descriptively, little is gained by adopting z-scores in terms of how handedness might be quantified on a continuous scale of measurement. Moreover, this example leads to another question: Would one prefer to have many observations on fewer subjects (i.e. more subjects like chimpanzee two in our example) or fewer observations on many more subjects (i.e. more subjects like chimpanzee one in our example)? Ultimately, it boils down to a tradeoff between internal and external validity in the research. Ideally, it would be nice to find better balance between these two research goals because, as it stands, scientists working on handedness in non-humans are opting for greater internal validity whereas individuals studying with humans are trending toward greater external validity.

There are several limitations to the studies on manual gestures and handedness that limits the generalization of our results. First, we did not make a strong distinguish between different types of specific gestures but rather focused on rather broad distinction in gestures and this could influence hand preferences (see Figure 2). Our analyses similarly suggest that differences in hand preference may be evident within and between species depending on the type of gesture. Unfortunately, the sample sizes were very small for some types of gestures and some caution should be exercised in interpreting these findings. Thus, it would be prudent in future studies to attempt to examine hand preference as a function of specific gesture types in which many more observations are available in each subject.

Second, this paper focused on manual asymmetries for gestures in great apes; however, additional studies on asymmetries in monkeys would be of interest. For example, as noted above, Meguerditchian and Vauclair (2006) reported significant population-level right handedness in HAND SLAPPING by baboons. HAND SLAPPING is an intra- and interspecific threat gesture produced by baboons and, in many ways, is similar to the HIT/SLAP behavior described in the present chapter for the apes. Thus, there seems to be some similar patterns of results between baboons and apes when considering handedness for gestures that are motorically and functionally similar.

In summary, the results of this meta-analysis on the existing data on hand preference for manual gestures in great apes suggest that population-level asymmetries are present, particularly among chimpanzees, bonobos and gorillas. The results are less clear in orangutans. For all apes, there is a need for additional data not just from captive individuals, but particularly from apes living in their natural environments. It is somewhat surprising that there are no data from wild apes given the historical documentation and interest in gestural communication within the context of different theories on the evolution of language. Moreover, with the evidence of an association between gestural communication and variation in both the anatomy and function of the Broca's area homolog in the chimpanzee brain (Taglialatela, Cantalupo & Hopkins 2006; Taglialatela, Russell, Schaeffer & Hopkins 2008), it seems that studies on lateralization in species-typical gestural communication might be an important and fruitful area for comparative studies on the evolution of brain and behavior.

Acknowledgement

This research was supported in part by NIH grants NS-42867, NS-73134, HD-60563, and HD-56232 as well as NIH grant U42 to the M.D. Anderson Cancer Center. American

Psychological Association guidelines for the ethical treatment of animals were adhered to during all aspects of this study.

References

Andrew, Richard J. 1991. "The nature of behavioral lateralization in the chick." In *Neural and Behavioral Plasticity. The use of the chick as a model,* Richard. J. Andrew (ed.), 536–554. Oxford: Oxford University Press.

Annett, Marian. 2002. *Handedness and Brain Asymmetry: The right shift theory.* Hove: Psychology Press.

Arbib, Michael. 2005. "From monkey-like action recognition to human language: An evolutionary framework for neurolinguistics." *Behavioral and Brain Sciences* 28 (2): 105–167.

Binder, Jeffrey R., Frost, Julie A., Hammeke, Thomas A., Cox, Robert W., Rao, Steven M. and Prieto, Thomas. 1997. "Human brain language areas identified by functional magnetic resonance imaging." *The Journal of Neuroscience* 17 (1): 353–362.

Biro, Dora, Inoue-Nakamura, Noriko, Tonooka, Rikako, Yamakoshi, Gen, Sousa, Claudia and Matsuzawa, Tetsuro. 2003. "Cultural innovation and transmission of tool use in wild chimpanzees: Evidence from field experiments." *Animal Cognition* 6 (4): 213–223.

Biro, Dora, Sousa, Claudia and Matsuzawa, Tetsoro. (2006). "Ontogeny and cultural propagation of tool use by wild chimpanzees at Bossou, Guinea: Case studies in nut cracking and leaf folding." In *Cognitive Development of Chimpanzees,* Tetsuro Matsuzawa, Masaki Tomonaga and Masayuki Tanaka (eds), 476–507. New York: Springer.

Bisazza, Angelo, Cantalupo, Claudio, Robins, Angela, Rogers, Lesley J. and Vallortigara, Giorgio. 1996. "Right-pawedness in toads." *Nature* 379 (6564): 408.

Boesch, Christophe. 1991. "Handedness in wild chimpanzees." *International Journal of Primatology* 12 (6): 541–558.

Bradshaw, John L. 1997. *Human Evolution: A neuropsychological perspective.* Hove, U.K.: Psychology Press.

Bradshaw, John L. and Rogers, Lesley J. 1993. *The Evolution of Lateral Asymmetries, Language, Tool Use, and Intellect.* San Diego: Academic Press.

Byrne, Richard W. and Byrne, Jennifer M. 1991. "Hand preferences in the skilled gathering tasks of mountain gorillas (*Gorilla gorilla berengei*)." *Cortex* 27: 521–536.

Call, Josep and Tomasello, Michael. 2007. *The Gestural Communication of Apes and Monkeys.* New York: Lawrence Erlbaum Associates.

Cartmill, Erica and Byrne, Richard W. 2007. "Orangutans modify their gestural signaling according to their audience's comprehension." *Current Biology* 17 (15): 1–14.

Cartmill, Erica. A and Byrne, Richard W. 2010. "Semantics of primate gestures: Intentional meanings of orangutan gestures." *Animal Cognition* 13 (6): 793–804.

Cashmore, Lisa, Uomini, Natalie and Chapelain, Amandine. 2008. "The evolution of handedness in humans and great apes: A review and current issues." *Journal of Anthropological Studies* 86: 7–35.

Corballis, Michael. C. 1992. *The Lopsided Brain: Evolution of the generative mind.* New York: Oxford University Press.

Corballis, Michael. C. 2002. *From Hand to Mouth: The origins of language.* Princeton, NJ: Princeton University Press.

Corp, Natalie and Byrne, Richard W. 2004. "Sex difference in chimpanzee handedness." *American Journal of Physical Anthropology* 123 (1): 62–68.
Cochet, Helene and Vauclair, Jacques. 2010. "Pointing gestures produced by toddlers from 15 to 30 months: Different functions, hand shapes and laterality patterns." *Infant Behavior and Development* 33 (4): 431–441.
Crow, Timothy. 2004. "Directional asymmetry is the key to the origin of modern Homo sapiens (the Broca-Annett axiom): A reply to Rogers' review of The Speciation of Modern Homo Sapiens." *Laterality: Asymmetries of Body, Brain and Cognition* 9 (2): 233–242.
Fletcher, Alison. W. 2006. "Clapping in chimpanzees: Evidence of exclusive hand preference for a spontaneous, bimanual gesture." *American Journal of Primatology* 68 (11): 1081–1088.
Fletcher, Alison W. and Weghorst, Jennifer A. (2005). "Laterality of hand function in naturalistically housed chimpanzees (*Pan troglodytes*)." *Laterality* 10 (3): 219–242.
Genty, Emilie, Breuer, Thomas, Hobaiter, Catherine and Byrne, Richard W. 2009. "Gestural communication of the gorilla (*Gorilla gorilla*): Repertoire, intentionality, and possible origins." *Animal Cognition* 12 (3): 527–546.
Güntürkün, Onur. 1997. "Avian visual lateralization: A review." *Neuroreport* 8 (6): 3–11.
Harrison, Rebecca M. and Nystrom, Pia. 2008. "Handedness in captive bonobos (*Pan paniscus*)." *Folia Primatologica* 79 (5): 253–268.
Harrison, Rebecca M. and Nystrom, Pia. 2010. "Handedness in captive gorillas." *Primates* 51 (3): 251–261.
Hatta, Takesha. 2006. "Handedness and the brain: A review of brain-imaging techniques." *Magnetic Resonance in Medical Science* 6 (2): 99–112.
Hewes, Gordon W. 1976. "The current status of the gestural theory of language origin." *Annals of the New York Academy of Sciences* 280: 482–504.
Hickok, Gregory, Bellugi, Ursula and Klima, Edward S. 1998. "The neural organization of language: Evidence from sign language aphasia." *Trends in Cognitive Sciences* 2 (4): 129–136.
Hopkins, William D. 1999. "On the other hand: Statistical issues in the assessment and interpretation of hand preference data in non-human primates." *International Journal of Primatology* 20 (6): 851–866.
Hopkins, William D. 2006. "Comparative and familial analysis of handedness in great apes." *Psychological Bulletin* 132 (4): 538–559.
Hopkins, William D. and Cantalupo, Claudio. 2005. "Individual and setting differences in the hand preferences of chimpanzees (*Pan troglodytes*): A critical analysis and some alternative explanations." *Laterality* 10 (1): 65–80.
Hopkins, William D. and Nir, Talia. 2010. "Planum temporale surface area and grey matter asymmetries in chimpanzees (*Pan troglodytes*): The effect of handedness and comparison with findings in humans." *Behavioural Brain Research* 208 (2): 436–443.
Hopkins, William D., Phillips, Kimberly A., Bania, Amanda, Calcutt, Sarah E., Gardner, Molly, Russell, Jamie, et al. 2011. "Hand preferences for coordinated bimanual actions in 777 great apes: Implications for the evolution of handedness in Hominis." *Journal of Human Evolution* 60: 605–611.
Hopkins, William D., Russell, Jamie L., Freeman, Hani, Buehler, Natalie, Reynolds, Elizabeth and Schapiro, Steven J. 2005. "The distribution and development of handedness for manual gestures in captive chimpanzees (*Pan troglodytes*)." *Psychological Science* 16 (6): 487–493.

Hopkins, William D., Russell, Jamie L., Lambeth, Susan and Schapiro, Steven J. 2007. "Handedness and neuroanatomical asymmetries in captive chimpanzees: A summary of 15 years of research". In *Evolution of Hemispheric Specialization in Primates,* William D. Hopkins (ed.): 112–135. London: Academic Press.

Hopkins, William D., Stoinski, Tara, Lukas, Kristen, Ross, Stephen and Wesley, Michael J. 2003. "Comparative assessment of handedness for a coordinated bimanual task in chimpanzees (*Pan*), gorillas (*Gorilla*), and orangutans (*Pongo*)." *Journal of Comparative Psychology* 117 (3): 302–308.

Horwitz, Barry, Amunts, Katrin, Bhattacharyya, Rajan, Patkin, Debra, Jeffries, Keith, Zilles, Karl and Braun, Allen R. 2003. "Activation of Broca's area during the production of spoken and signed language: a combined cytoarchitectonic mapping and PET analysis." *Neuropsychologia* 41 (14): 1868–1876.

Humle, Tatiana and Matsuzawa, Tetsuro. 2009. "Laterality in hand use across four tool use behaviors among the wild chimpanzees of Bossou, Guinea, West Africa." *American Journal of Primatology* 71 (1): 40–48.

Kalan, Ammie K. and Rainey, Hugo J. 2009. Hand-clapping as a communicative gesture by wild female swamp gorillas. *Primates* 50: 273–275.

Kimura, Doreen. 1993. *Neuromotor Mechanisms in Human Communication.* New York: Clarendon Press/Oxford University.

Knecht, Stefan, Deppe, Michael, Drager, Bianca, Bobe, Linus, Lohmann, Hubertus, Ringelstein, Ellen B., et al. 2000. "Language lateralization in healthy right-handers." *Brain* 123 (1): 74–81.

Knecht, Stefan, Drager, Bianca, Deppe, Michael, Bobe, Linus, Lohmann, Hubertus, Floel, Agnes, et al. 2000. "Handedness and hemispheric language dominance in healthy humans." *Brain* 123 (12): 2512–2518.

Leavens, David A., Hostetter, Autumn B., Wesley, Michael J. and Hopkins, William D. 2004. "Tactical use of unimodal and bimodal communication by chimpanzees, *Pan troglodytes*." *Animal Behaviour* 67 (3): 467–476.

Leavens, David A., Russell, Jamie L. and Hopkins, William D. 2005. "Intentionality as measured in the persistence and elaboration of communication by chimpanzees (*Pan troglodytes*)." *Child Development* 76 (1): 291–306.

Liebal, Katja, Call, Josep and Tomasello, Michael. 2004. "Use of gesture sequences in chimpanzees." *American Journal of Primatology* 64 (4): 377–396.

Liebal, Katja, Müller, Cornelia and Pika, Simone. (eds). 2005. *Gestural Communication in Nonhuman and Human Primates.* Amsterdam/Philadelphia: John Benjamins Publishing Company.

Liebal, Katja, Pika, Simone, Call, Josep and Tomasello, Michael. 2004. "To move or not to move: how apes adjust to the attentional state of others." *Interaction Studies* 5 (2): 199–219.

Liebal, Katja, Pika, Simone and Tomasello, Michael. 2006. Gestural communication in orangutans (*Pongo pygmaeus*). *Gesture* 6 (1): 1–38.

Lonsdorf, Elizabeth V. and Hopkins, William D. 2005. "Wild chimpanzees show population level handedness for tool use." *Proceedings of the National Academy of Sciences,* 102 (35): 12634–12638.

Marchant, Linda F. and McGrew, William C. 2007. "Ant fishing by wild chimpanzees is not lateralised." *Primates,* 48 (1): 22–26.

McGrew, William C. and Marchant, Linda F. 1997. "On the other hand: Current issues in and meta-analysis of the behavioral laterality of hand function in non-human primates." *Yearbook of Physical Anthropology* 40: 201–232.

Meguerditchian, Adrien, Calcutt, Sarah E., Lonsdorf, Elizabeth V., Ross, Stephen R. and Hopkins, William D. 2010. "Captive chimpanzees are right-handed for bimanual feeding." *American Journal of Physical Anthropology* 141 (4): 638–645.

Meguerditchian, Adrien and Vauclair, Jacques. 2006. "Baboons communicate with their right hand." *Behavioural Brain Research* 171 (1): 170–174.

Meguerditchian, Adrien, Vauclair, Jacques and Hopkins, William D. 2010. "Captive chimpanzees use their right hand to communicate with each other: Implications for the origins of hemispheric specialization for language." *Cortex* 46 (1): 40–48.

Palmer, Alan R. 2002. "Chimpanzee right-handedness reconsidered: Evaluating the evidence with funnel plots." *American Journal of Physical Anthropology* 118: 191–199.

Peters, Helen H. and Rogers, Lesley J. 2007. "Limb use and preferences in wild orangutans during feeding and locomotor behavior." *American Journal of Primatology* 70 (3): 261–270.

Pika, Simone, Liebal, Katja and Tomasello, Michael. 2003. "Gestural communication in young gorillas (*Gorilla gorilla*): Gestural repertoire, learning, and use." *American Journal of Primatology* 60 (3): 95–111.

Pika, Simone, Liebal, Katja and Tomasello, Michael. 2005. "The gestural repertoire of bonobos (*Pan paniscus*): Flexibility and use." *American Journal of Primatology* 65 (1): 39–61.

Pollick, Amy S. and de Waal, Frans M. B. 2007. "Ape gestures and language evolution." *Proceedings of the National Academy of Sciences* 104 (19): 8184–8189.

Porac, Claire and Coren, Stanley. 1981. *Lateral Preferences and Human Behavior*. New York: Springer.

Poss, Sarah R., Kuhar, Chris, Stoinski, Tara S. and Hopkins, William D. 2006. "Differential use of attentional and visual communicative signaling by orangutans (*Pongo pygmaeus*) and gorillas (*Gorilla gorilla*) in response to the attentional status of a human." *American Journal of Primatology*, 68 (10): 978–992.

Rasmussen, Thomas and Milner, Brenda. 1977. "The role of early left-brain injury in determining lateralization of cerebral speech function." *Annals of the New York Academy of Science* 299: 355–369.

Rizzolatti, Giacomo and Arbib, Michael A. 1998. "Language within our grasp." *Trends in Neuroscience* 21 (5): 188–194.

Rogers, Lesley J. and Andrew, Richard J. 2002. *Comparative Vertebrate Lateralization*. Cambridge: Cambridge University Press.

Rogers, Lesley J. and Kaplan, Gisela. 1995. "Hand preferences and other lateral biases in rehabilitated orangutans, *Pongo pygmaeus pygmaeus*." *Animal Behaviour* 51 (1): 13–25.

Shafer, Doreen D. 1993. "Patterns of hand preference in gorillas and children." In *Primate Laterality: Current Behavioral Evidence of Primate Asymmetries*, Jeannette P. Ward and William D. Hopkins (eds), 267–283. New York: Springer-Verlag.

Shafer, Doreen D. 1997. "Hand preference behaviors shared by two groups of captive bonobos." *Primates* 38 (3): 303–313.

Taglialatela, Jared P., Cantalupo, Claudio and Hopkins, William D. 2006. "Gesture handedness predicts asymmetry in the chimpanzee inferior frontal gyrus." *NeuroReport* 17 (9): 923–927.

Taglialatela, Jared P., Russell, Jamie L., Schaeffer, Jennifer A. and Hopkins, William D. 2008. "Communicative signaling activates "Broca's" homologue in chimpanzees." *Current Biology* 18 (5): 343–348.

Toga, Arthur W. and Thompson, Michael. 2003. “Mapping brain asymmetry.” *Nature Reviews Neuroscience* 4: 37–48.
Vallortigara, Giorgio and Rogers, Lesley J. 2005. “Survival with an asymmetrical brain: Advantages and disadvantages of cerebral lateralization.” *Behavioral and Brain Sciences* 28: 574.
Warren, James M. 1980. “Handedness and laterality in humans and other animals.” *Physiological Psychology* 8 (3): 351–359.

Mandrill visual gestures

A round-the-world study of the largest of all monkeys

Mark E. Laidre
Department of Integrative Biology, University of California, Berkeley

Apes have rich gestural communication systems. But do monkeys? If so, what forms and functions do monkey gestures have, and how does the relationship between form and function originate developmentally? This chapter reports intensive observations aimed at answering these questions in the largest of all monkeys, the mandrill (*Mandrillus sphinx*). Nearly thirty captive mandrill groups distributed across three continents were studied over the course of a decade, some groups for three consecutive generations. Mandrills performed three visual gestures: SLAP GROUND, HAND EXTENSION, and EYE COVERING. While SLAP GROUND was distributed uniformly across individuals and groups, HAND EXTENSION and EYE COVERING – the two focal gestures of this chapter – showed restricted distributions that were consistent with their being learned rather than instinctive. Contrasting HAND EXTENSION against EYE COVERING for fifteen different signal parameters revealed that these two gestures had antithetical forms and functions but had similar developmental histories. Within each gesture the close correspondence between form and function suggested that ontogenetic ritualization contributed to both HAND EXTENSION's and EYE COVERING's origin. Furthermore, for EYE COVERING the clustering of all known performers within a single group suggested this gesture was culturally transmitted. Broadly, these results suggest that multiple interacting factors, including instinct, learning, and culture may contribute to the emergence of monkey gestures. Follow-up experiments, however, remain critical for rigorously testing the developmental origins and potential cultural transmission of non-human gestures.

1. Introduction

Animals communicate through diverse means, encompassing chemical, electric, acoustic, optical, and tactile modalities (Bradbury & Vehrencamp 2011). In *King Solomon's Ring* (1952), one of the classic works in animal communication, Konrad

Lorenz focused on communication via sight and touch, discussing at length a special form of communication which he termed "gesture" (p. 43). Although Lorenz never formally defined gesture he attributed gestures to a wide array of animals, including fish, jackdaws, eagles, turkeys, wolves, dogs, and humans (Lorenz 1952). Surprisingly, despite many references to apes and monkeys, Lorenz made no mention in *King Solomon's Ring* of the existence of gestures in non-human primates, the taxonomic group with the most well developed manual dexterity. Recently, however, the study of simian gestures has developed into a burgeoning field (Smuts 2002; de Waal 2003; Call & Tomasello 2007; Liebal et al. 2007; Pika & Liebal this volume). While a universal definition of gesture may still be elusive (Struhsaker 1967; Scott & Pika this volume), there is nevertheless broad appreciation that the gestures of our primate relatives have much to tell us about the intricacies of communication.

A foundational question in the study of primate gestural communication is how gestures first come into being during the lifetime of an organism. Like any behavioral or morphological phenotype, gestures must arise developmentally based on some interplay of genes and environment (Dawkins 1976). In theory, three possibilities exist by which gestures could be acquired during ontogeny. First, gestures could be acquired innately, being passed down through raw genetic inheritance, with little or no feedback from the environment. In such cases, gestures should be similar between individuals within the same group and across different groups; also, as long as the population has a uniform gene pool, gestures should be passed down unchanged across generations (Lorenz 1952). Second, gestures could be acquired through individual trial-and-error learning, as organisms repeatedly interact with their environment and these interactions shape the types of gestures that eventually emerge. In such cases, gestures should frequently be idiosyncratic due to the unique learning trajectories taken by different individuals, and so gestures should vary between individuals within the same group, across different groups, and across generations (Tomasello et al. 1997). Finally, gestures could be acquired through some form of social learning and cultural transmission, such as imitation, in which individuals copy the gestures they observe other community members performing. In such cases, gestures should be similar between individuals in the same group, who can all learn from one another, but different between individuals in separate groups, who lack the opportunity to learn from one another (de Waal 2003; Laland & Galef 2009). It is noteworthy that substantial debate exists over whether animals possess 'culture'. Yet in some non-human species, individuals might possess a capacity to copy behaviors of other group members, which could cause the behavior of different groups to diverge (Laland & Galef 2009). As defined in a minimalist sense (Bonner 1980), animal culture therefore exists when socially-learned behavior is group-specific, such that a behavior is performed by multiple individuals in a group but is absent from other groups of the same species.

Critically, these three explanations for gestural acquisition need not be mutually exclusive. Not only could different gestures in a given species' repertoire be acquired by each of the three aforementioned modes, but all three of these modes could even jointly contribute to the development of single gestures. Which explanation, or combination of explanations, is most likely to account for the way non-human primates acquire their gestures? Some of the earliest detailed studies of gestural communication in non-human primates by Tomasello and colleagues provided support for the second explanation, involving individual trial-and-error learning. Based on longitudinal observations of a group of captive chimpanzees, Tomasello et al. (1997) argued that a process of 'ontogenetic ritualization' – invoking repeated interactions and social feedback – best accounted for how great apes acquire the majority of their gestures. Subsequent studies on other great apes showed that social learning, however, plays a crucial role for the acquisition of some gestures (Pika et al. 2005). More recently, studies of captive gorillas (Genty et al. 2009) as well as wild chimpanzees (Hobaiter & Byrne 2011) have suggested that non-human primates might inherit the majority of their gestures based on the first explanation, a hard-wired genetic mechanism. Given the possible importance of multiple modes of gestural acquisition, it is notable that relatively little research has focused on the third mode, social learning and culture. Indeed, most studies of primate gestural communication have involved observations of just one or, at most, a few groups of the same species (Call & Tomasello 2007), thus limiting opportunities to investigate diversity between groups and to test the predictions outlined above. Beyond studying multiple groups, what settings and methods would be most conducive to identifying social learning and cultural transmission of gestures in non-human primates?

Studies of wild primates have certainly contributed to our knowledge of cultural gestures (Whiten et al. 1999), and ongoing studies in the wild continue to provide intriguing data on primate gestures (e.g. referential gestures, Pika & Mitani 2006). However, even though wild studies may be superior to captive ones in certain respects, like ecological validity (Boesch 2007), for researchers exploring cultural gestures and developmental origins in general, studies of captive primates provide numerous advantages over studies in the wild. Different captive groups are effectively separate, allopatric populations with non-overlapping ranges, which enables the gestural behavior of different groups to be readily contrasted. And since groups are often housed in very similar, sometimes identically-controlled environments, the confound of ecological variation is minimized while the opportunity for close-range observation is maximized. Rates of social innovation are also greater in captive than in wild primates (Kummer & Goodall 1985), providing researchers with better chances of observing the initial innovation of a gesture and monitoring whether and how the gesture subsequently spreads within the group. Furthermore, different captive groups only come into contact during discrete and usually planned translocations in which individuals from one community are transferred to another to maintain genetic uniformity in the population. In cases where the translocated individual is one that performs a unique gesture, it is possible to rigorously examine social learning and cultural diffusion of the gesture within a naïve community. Finally, compared to the wild, captive settings enable superior opportunities for manipulating variables and conducting novel experiments, like video playbacks, which can complement observation-based research. Thus, for investigating the development of gestures, especially cultural acquisition, between-group comparisons are useful.

The present chapter reports a study of gestural communication and culture across nearly thirty captive monkey groups. This study has lasted just short of a decade and, for some groups, has involved longitudinal observations extending over three generations. To my knowledge it is the most geographically-extensive and long-term investigation of gestural communication in any monkey species. The species that is the focus of this work is the mandrill (*Mandrillus sphinx*), the most brightly colored of all mammals (Darwin 1871). The mandrill's bright coloration underscores a rich communication system based on vision, the modality that forms the core of gestural signals. This chapter focuses specifically on visual-based manual gestures, defined here as actions involving the freely-moving hands and arms, which visually communicate information to conspecifics and influence their behavior (cf. Smuts 2002; de Waal 2003; for other definitions see Scott & Pika this volume). Manual acts involving tool-use (Whiten et al. 1999) and also some gestures (de Waal 2003) have yielded evidence for learning and cultural variation in apes. An objective of the present investigation was to contrast the manual gestures of different mandrill groups to identify potential cultural variants in monkeys. Two mandrill gestures, in particular, which have recently been analyzed in detail, are contrasted in this paper: HAND EXTENSION (Laidre 2008) and EYE COVERING (Laidre 2011). The results section partly draws on these two prior publications, occasionally referring to some of the figures in these papers. By contrasting these two gestures, this chapter seeks to show the divergent routes that gestures can take in their form and function in separate groups, which is a critical goal in animal culture (Krützen et al. 2007). Another objective was to examine the distribution of the gestures within and across study groups to generate insights into how these gestures may have originated developmentally and how each gesture's form-function link was forged. Finally, a critical component of the overall research program described herein is to follow up the long-term observational research with experiments that test plausible developmental origins scenarios, especially scenarios invoking cultural transmission.

2. Methods

2.1 Study system: The mandrill in captivity

Mandrills are endemic to the tropical rain forests of equatorial west Africa. In the wild they occupy a small range that extends from Cameroon in the north to Gabon and the Congo in the south. Humans, however, have artificially increased the mandrills' range to a massive extent, creating captive groups around the world. Such captive groups provide useful observation conditions compared to the mandrills' densely-forested wild habitat, where it has been difficult to habituate free-ranging groups for detailed behavioral observation (Abernethy et al. 2002).

In this study, captive mandrill groups were located through ISIS, the 'International Species Information System' (www.isis.org), a continuously-updated database of animal species held in captivity at zoos and other institutions around the world. At the time of writing (1 Oct 2010) ISIS reported that worldwide there exist 560 captive mandrills distributed across 98 groups on six of the seven continents, excluding only Antarctica. Most captive mandrills are concentrated in Europe (71.8%) and North America (20.5%), with a small proportion (7.7%) also being found across Australia, Central and South America, Africa, and Asia. In addition, though not listed in the ISIS system, several hundred captive mandrills are contained in semi-free-ranging colonies in Gabon, West Africa at the Centre International de Recherches Médicales (personal observation).

2.2 Study groups

From 2001 to present, I studied twenty-seven mandrill groups across North America, Africa, and Europe. Figure 1A shows the locations of the study sites where these groups were observed. For detailed information on the sites, the composition of each group, the layout of each group's enclosure, the dates of observation, and the total hours of observation see Laidre (2008, 2009a, 2011). The mean observation time per group exceeded 50 h and altogether over 200 mandrills were studied. Dominance was assigned in each group by coding approach-avoid interactions in which one individual's approach caused the other to avoid. Three groups (Table 1) are focused on here, since unique gestures were performed within them involving a minimum of two individuals.

Figure 1. (A) Map of study sites for mandrill groups across three continents (North America, Africa, and Europe). Pictures of the gesture (B) HAND EXTENSION and (C) EYE COVERING

Table 1. Compositions of the two groups (Syracuse and Buffalo) where HAND EXTENSION was performed and the one group (Colchester) where EYE COVERING was performed. Names of gesturers are denoted in bold. Individuals are listed within each group in descending dominance order. Dashes indicate offspring of the above female.

Name (sex, age at start of study, date of birth)
SYRACUSE (N = 4)
JJ (, 13 yrs, born 6 Feb 1988); died 30 Dec 2008
Zenani (tç, 11 yrs, born 5 Feb 1990)
– **Kelley** (tç, 3 yrs, born 25 July 1998); transferred to Orlando summer 2010
– **Eebi** (tç, 4 yrs, born 30 March 1997)
BUFFALO (N = 4)
Savuti (, 10 yrs, born 30 Sept 1991); transferred in 20 Sept 2002
Gail (tç, 19 yrs, born 10 Jun 1983)
– **Valerie** (tç, 4 yrs, born 14 Feb 1998); transferred to Bronx Dec 2008
Louise (tç, 3 yrs, born 18 March 1999); transferred to Bronx Dec 2008
COLCHESTER (N = 23)
Dume (, 12 yrs, born 15 Sep 1994); transferred in 23 Jan 2004

(Continued)

Table 1. (Continued)

Name (sex, age at start of study, date of birth)
Celine (tç, 22 yrs, born 3 Jan 1985); transferred in 27 Oct 1988; died 22 May 2009
– **Kayin** (̧, 1 yr, born 13 Feb 2006)
Orinoko (tç, 13 yrs, born 8 Nov 1993)
– Olive (tç, under 1 yr, born 20 Dec 2006)
– Ollie (tç, 1 yr, born 10 Dec 2005)
– Obi (tç, 2 yrs, born 19 Sept 2004); transferred to Chester 23 Jan 2008
– Oakley (tç, 6 yrs, born 4 Feb 2001)
Malaya (tç, 23 yrs, born 14 Aug 1984); transferred in 27 Oct 1988
– Malik (̧, 2 yrs, born 3 Apr 2005)
– Matilde (tç, 6 yrs, born 21 Sep 2000)
– **Milly** (tç, 10 yrs, born 27 Sep 1996)
– Mathias (̧, under 1 yr, born 15 Nov 2006)
– Mosi (tç, 2 yrs, born 29 Oct 2004); transferred to Chester 23 Jan 2008
– Solomina (tç, 14 yrs, born 27 Sep 1992)
– Salem (̧, under 1 yr, born 24 Dec 2006)
– Nyah (tç, 1 yr, born 23 Nov 2005); transferred to Chester 23 Jan 2008
– Sefiya (tç, 2 yrs, born 20 Sep 2004); transferred to Chester 23 Jan 2008
Phoenix (̧, 10 yrs, born 12 Oct 1996); died 26 Jun 2008
Mac (̧, 9 yrs, born 18 May 1998)
Max (̧, 9 yrs, born 10 Jul 1998)
Barney (̧, 7 yrs, born 8 Oct 1999)
T.J. (̧, 5 yrs, born 22 Sep 2001)

2.3 Visual-based manual gestures

The major focus of mandrill between-group comparisons has been to investigate the species' social behavior and communication, including facial expressions (Laidre & Yorzinski 2005), postural displays (Laidre 2005), bodily movements not specialized for communication (Laidre 2009a), vocalizations (Laidre 2006), and overall social relationships (Laidre & Yorzinski 2008). The present chapter, however, focuses specifically on visual-based manual gestures. Any behaviors that were observed during the course of the study, and which fit the definition provided in the introduction, are considered here. For each gesture identified, I examine its form, its function (demarcated using the observational techniques expanded upon below), and the history of how it first emerged.

2.4 Observations and sampling regime

Data were collected both by focal animal sampling (Altmann 1974) and behavioral sampling (Martin & Bateson 1993). Using a microcassette recorder, I noted the exact time (to the nearest second) of each gestural onset and offset, as well as each related behavioral occurrence and surrounding context. An exhaustive listing of the contexts in which gestures could be embedded as well as the responses that recipients could exhibit toward gestures can be found in Laidre and Yorzinski (2005) and Laidre (2005, 2008, 2011). Together the contextual and response data were used as a basis for investigating each gesture's function: the contextual data revealed the type of situations where a gesture was most likely to occur (Laidre & Yorzinski 2005) and the sequence of behaviors performed by recipients before and after a gesture showed how the gesture influenced others' behavior (Laidre 2005).

3. Results

3.1 Manual gestures and distribution across study groups

Mandrills performed three visual-based, manual gestures: SLAP GROUND, HAND EXTENSION, and EYE COVERING (henceforth 'SG', 'HE', and 'EC'). SG, a well-established threat gesture (Struhsaker 1967), involved the hand being lifted perpendicularly from the ground and then being brought back in contact with the ground in an instantaneous pounding motion. SG was distributed uniformly, being observed across all groups, with three or more performers in each group (Laidre 2005). In contrast, HE was observed in only two groups (Syracuse, NY and Buffalo, NY, USA) and EC was observed in only one group (Colchester, England) (see Table 1).

3.2 HAND EXTENSION

HE involved the arm being raised approximately parallel to the ground and directed toward another individual, with the palm of the hand facing downwards and the outstretched arm remaining stationary for several seconds with no lateral movement (see Figure 1B). HE was first observed in 2002 being performed by all of the four immatures in both the Syracuse and the Buffalo group. The performers (see Table 1) encompassed two females aged three and four years in Syracuse and two females of the same ages in Buffalo. HE was directed exclusively at more dominant individuals in these groups, typically a performer's parent or an unrelated adult. HE's function appeared to be to provoke dominants: its probability of occurring in any context was highest (4.8%) in the prior to agonism context (see also Figure 2 in Laidre 2008), and of several communicative acts HE had the highest chance (21.6%) of eliciting a response in which the recipients' behavior changed from a non-aggressive state to an aggressive state (see also Figure 3 in Laidre 2008). Individuals tended to perform HE when their age mate was unavailable as a play partner, resorting in effect to aggravating their elders as an alternative social play outlet. Observations carried out in 2008 confirmed that HE was still being performed by the same individuals in both Buffalo and Syracuse, the performers then being all over nine yrs old. Since that time both the Buffalo and Syracuse groups have fractured due to a combination of deaths and translocations (see Table 1).

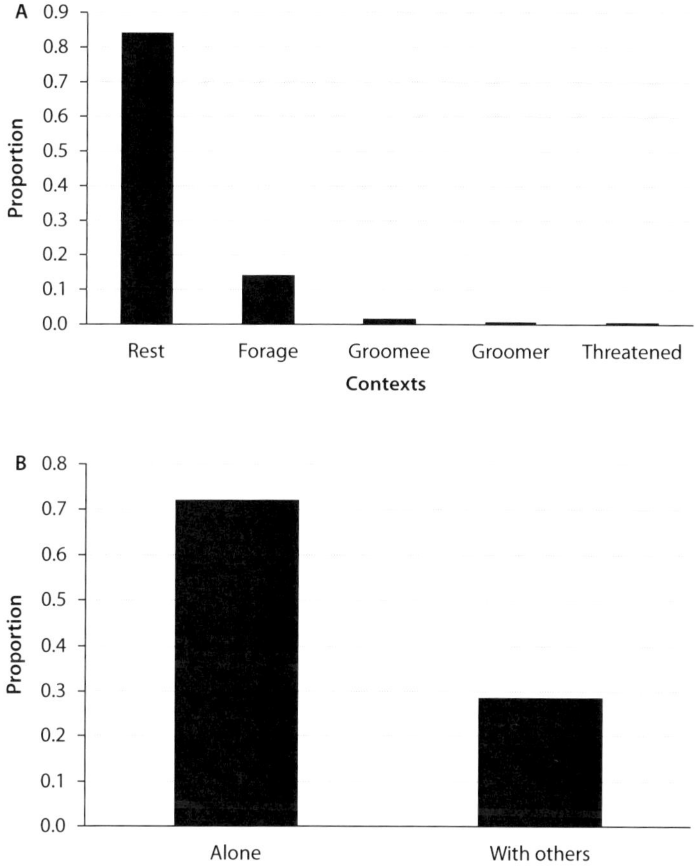

Figure 2. (**A**) Behavioral contexts in which EYE COVERING was performed. Based on a total of 815 (N = 815) gestural occurrences from focal and behavioral sampling combined. 'Rest': absence of socializing; 'Forage': identical to 'Rest' except picking through grass and eating; 'Groomee': being groomed by another; 'Groomer': grooming another; 'Threatened': chased or received a threat display moments before. (**B**) Whether at the time its gesture was first detected the gesturer was 'Alone' (i.e. no one else within 2 m; N = 520) or 'With others' (i.e. had other community members, aside from fully-dependent infants, within 2 m; N = 202). Data are derived from three years of observation (2007, 2008, and 2009)

3.3 EYE COVERING

EC involved one or both hands being brought overtop the face, covering the eyes, with the elbow often simultaneously raised prominently into the air (see

Figure 1C). EC was held for prolonged periods averaging 8.8 ± 1.5 min. Seven individuals, aged two to ten years, female and male performed EC in the Colchester group, and these performers tended to occupy the bottom of the dominance hierarchy among the twenty-three individuals in the group (see Table 1). EC was first observed being performed in 1999 by an immature female, then aged three years (see also Laidre 2011). This female continues to perform EC to the present. Thus altogether, EC has lasted within Colchester for more than a decade, enduring deaths, removals, and births. EC's function appeared to be like that of a 'do not disturb' sign: the gesture was performed primarily when individuals were alone, either resting or foraging (see Figure 2). If individuals performed EC, they received fewer approaches and fewer tactile contacts compared to control periods when they were not covering their eyes (for approaches: 5.10 vs. 17.36 per h; for touches: 0.39 vs. 1.54 per h; see also Figure 3 in Laidre 2011).

3.4 Contrast of HAND EXTENSION and EYE COVERING

A comparison of HE and EC across fifteen signal parameters showed these gestures were the antithesis of one another, especially with respect to parameters related to their forms and functions (see Table 2). HE whose function was to provoke dominants involved the distal most – and hence lowest-risk – body part, the hand, being extended outward at specific recipients for brief durations. In contrast, EC whose function was to reduce disturbances from dominants involved the hand being brought inward to cover the eyes, broadcasting to many recipients at once and over a much longer duration that a prominent body part (typically needed for face-to-face interactions) was concealed.

Table 2. Two gestures contrasted with respect to their form, function, and history

	HAND EXTENSION (Laidre 2008)	EYE COVERING (Laidre 2011)
FORM		
Hand position	Directed outward, touching nothing; only performed with one hand at a time	Directed inward, touching oneself; sometimes performed with two hands simultaneously
Limb position	Elbow roughly straight	Elbow acutely bent and often raised upward like a 'flag'
Body position	Poised to interact; usually standing, sometimes sitting	Reclined, lying or sitting
Temporal dynamics	Instantaneous (always < 10 sec)	Prolonged (often > 10 min)
Directedness	Aimed at single, specific recipients	Broadcast widely to an audience encompassing any potential recipients

(*Continued*)

Table 2. (Continued)

	HAND EXTENSION (Laidre 2008)	EYE COVERING (Laidre 2011)
FUNCTION		
Contexts of use	Social contexts, especially prior to agonism	Non-social contexts, especially during rest or foraging
Responses elicited	Immediate response; incites others	Long-term response; keeps others away
Meaning	'Bring it on'	'Do not disturb'
HISTORY		
First performer	~3 yr old female	~3 yr old female
Total performers	Four young females; two in each of two separate groups (Buffalo and Syracuse)	Seven individuals, including males and females, old and young individuals, all in one group (Colchester)
Rank of performers	Low ranking	Low ranking
Group size	Small (four individuals in each of two groups)	Large (23 individuals in one group)
Durability	≥ six years; more observations required to determine if still ongoing after group alterations	> ten years; still ongoing
Developmental origin	Ontogenetic ritualization; playful exploration while trying to minimize risk	Ontogenetic ritualization; attempt to hide by cutting off visual contact
Cultural transmission?	Possibly	More probable

HE and EC were not opposites in all respects though. Both gestures were performed by individuals of low rank (see Table 1), and the first individuals to perform these gestures (Eebi in Syracuse, Louise in Buffalo, and Milly in Colchester) were young immature females, approximately three years of age. The developmental time window of HE's and EC's origin thus coincided closely.

4. Discussion

While substantial research interest has centered on the gestures of apes (e.g. de Waal 2003; Pika et al. 2005; Pika & Liebal this volume), the study of monkey gestures has been relatively neglected (Maestripieri 1999; Smuts 2002; Hesler & Fischer 2007). Perhaps as a consequence, some researchers have been skeptical of whether monkeys possess gestures at all, with suggestions even being made that "one of the absolute differences between monkey and ape visual communication is the absence of free-hand gestures in monkeys" (de Waal 2003, p. 27). The species *Mandrillus sphinx* has been the focus of the present chapter. As the work reviewed here has shown, mandrills, like apes and humans, do in fact exhibit gestures, defined in a way (see introduction) that is broadly consistent with some other definitions of 'gesture' that have been proposed in the non-human primate literature (cf. Smuts 2002; Pollick & de Waal 2007). The number of gestures identified in monkeys, however, is several times smaller than the number identified in apes (Call & Tomasello 2007).

Beyond documenting the existence of monkey gestures, and examining their form and function, an important goal of research on mandrill gestures has been to contribute to our understanding of how non-human gestures originate developmentally. Thus far, this objective has been pursued indirectly, by examining which individuals in which groups perform specific manual gestures. In the discussion that follows, I consider what the observational research carried out on mandrills over the past decade tells us about gestural development, and I outline experiments that are critical to rigorously testing developmental-origins scenarios like cultural transmission.

4.1 Developmental origins

Of the mandrills' gestures, SG was the only one that was distributed uniformly across individuals and groups, consistent with the first developmental explanation, invoking an innate genetic tendency. SG is also found in the repertoires of many other species of Old World monkey (Struhsaker 1967), so it appears to be an ancient, hard-wired element of monkey communication. In contrast to SG, the mandrills' two other visual gestures (HE and EC) were more restricted in their distributions across individuals and groups, suggesting that learning and not mere genetic hard-wiring played some role in their development.

Interestingly, in comparison with one another, HE and EC had virtually opposite forms and opposite functions. Individually though each gesture showed a close internal correspondence between its own form and function. This form-function correspondence suggests that a process of 'ontogenetic ritualization' (Tomasello et al. 1997) contributed to the creation of each gesture. In this process an incipient movement, which possesses no inherent meaning to begin with, can be shaped into a meaningful gesture over the course of repeated interactions. This happens as recipients gradually recognize that the precursor can reliably predict the actor's mood or intentions (Tomasello et al. 1997).

For HE, the ontogenetic ritualization process seemed to build on an extant gesture, SG, which is a threat display that immature mandrills commonly perform toward adults to test the dominance hierarchy in a playful, explorative way. Unlike the mandrills' other threat displays (which entail bringing vulnerable body parts like the head and eyes toward an opponent; Laidre 2005), SG involves a reaching-out, pawing-like movement that is of lower risk. Given how often immatures performed SG (Laidre 2008), this gesture could have served as a foundation for the ritualization of HE, which is an even less-risky manual act to use for inciting dominants.

For EC, ontogenetic ritualization may have also contributed to the gesture's origin. One plausible developmental scenario implicates limitations in monkeys' ability to assume others' visual perspective. Although monkeys appear to possess some sensitivity to what others perceive (Flombaum & Santos 2005), they are generally thought to have a limited 'theory-of-mind' capacity (Cheney & Seyfarth 2008; Laidre 2006), comparable to human children under a few years in age (Tomasello & Call 1997). Children this young typically do not understand that others' knowledge may differ from their own depending on perspective, and as a consequence children sometimes incorrectly assume that by covering their eyes they can prevent others from seeing them (Tomasello & Call 1997). The origin of EC may have been based on this 'mistake', with the gesture first being enacted when an individual sought seclusion and was attempting to hide itself from others' view. Being disturbed in this superficially 'hidden' state would have been particularly aversive, prompting a strong avoidance response in the actor. And this response, repeated over many interactions, would have impressed upon disturbers the reduced social interest an individual had when performing EC.

Finally, there was no evidence that either gesture was in any way shaped by human intervention (see Laidre 2011). Indeed, both gestures were performed irrespective of whether humans were in the vicinity of the mandrills' exhibit, and these gestures have not been deployed in any human-geared contexts, as while being fed.

4.2 Cultural transmission

Some of the earliest evidence for animal culture derived from monkeys' manual acts (Kawamura 1959), in particular Japanese macaques' washing of potatoes in the sea. This evidence has been criticized, however, since different monkeys might plausibly have learned the washing technique independently, without any social learning involved (Laland & Galef 2009). Could the same be true for the mandrill gestures reported herein? For HE the evidence for cultural transmission was not strong: only a total of four individuals performed HE, two in each of two separate groups. At most there could have been one case of social learning in each group; or, alternatively, a second instance of individual innovation in each group. In contrast, for EC the evidence for cultural transmission was stronger, albeit indirect. All seven known performers of EC clustered into a single group. Moreover, alternative hypotheses to social learning for EC's unique clustering were not supported: genetic variation, habitat differences, social composition differences, inadequate sampling, multiple independent innovations, human inducement, and training were each unsupported (Laidre 2011). Nevertheless, for cultural transmission to be rigorously established for EC, or for any behavior, experiments are critical.

Recent experiments have explored extensively the cultural transmission of foraging and tool-use in non-humans (reviewed in Whiten & Mesoudi 2008), but few experiments to date have explored the cultural transmission of gestures (Tomasello et al. 1997). The question of whether non-humans culturally transmit gestures remains unresolved, perhaps because a critical experiment – transplanting individuals with novel gestures to new groups – is logistically challenging (Laland & Hoppitt 2003). One alternative means by which the cultural transmission of gestures can be explored is video playback. Paukner and Anderson (2006) showed that video playbacks of conspecific yawns to macaque monkeys induce yawning in watching monkeys. Unlike learned manual gestures, yawns are an innate part of these monkeys' repertoire. But in principle the same experiment could be employed with a novel behavior like EC. Video playbacks of EC are planned at the San Francisco Zoo, a group where EC has never been observed. The playbacks will determine how a naïve community responds to a novel gesture and whether extended playback causes novice mandrills to adopt the gesture. While evidence for imitation of novel manual acts has not been found in non-human primates, monkeys do recognize when they are being mimicked in their social behavior (Paukner et al. 2009; Call & Carpenter 2009), and such a capacity could provide a foundation for monkeys to mimic gestures.

Acknowledgements

Thomas Seeley and Sandra Vehrencamp, my undergraduate advisors, provided a continuous source of encouragement from the time when this study began to the present. I am grateful to the zoos, field sites, and other institutions that allowed me to study their mandrills, including the universities that provided an intellectual home (Cornell during undergrad, Cambridge during masters, Princeton during doctorate, and Berkeley during postdoc). I thank Reyhan Furman for discussion about gestures and the workshop organizers (Simone Pika and Katja Liebal) for inviting me to the conference in Germany.

References

Abernethy, K.A., White, L. J. T. and Wickings, E. J. 2002. "Hordes of mandrills (*Mandrillus sphinx*): Extreme group size and seasonal male presence." *Journal of Zoology* 258: 131–137.

Altmann, Jeanne. 1974. "Observational study of behavior: Sampling methods." *Behaviour* 49: 227–267.

Boesch, Christophe. 2007. "What makes us human (*Homo sapiens*)? The challenge of cognitive cross-species comparison." *Journal of Comparative Psychology* 121: 227–240.
Bonner, John T. 1980. *The evolution of culture in animals*. Princeton: Princeton University Press.
Bradbury, Jack W. and Vehrencamp, Sandra L. 2011. *Principles of animal communication*. 2nd edition. Sunderland: Sinauer Associates.
Call, Josep and Carpenter, Malinda. 2009. "Monkeys like mimics." *Science* 325: 824–825.
Call, Josep and Tomasello, Michael (eds). 2007. *The gestural communication of apes and monkeys*. Mahwah: Lawrence Erlbaum Associates.
Cheney, Dorothy L. and Seyfarth, Robert M. 2008. *Baboon metaphysics*. Chicago: University of Chicago Press.
Darwin, Charles. 1871. *The descent of man and selection in relation to sex*. London: John Murray.
Dawkins, Richard. 1976. *The selfish gene*. Oxford: Oxford University Press.
de Waal, Frans B. M. 2003. "Darwin's legacy and the study of primate visual communication." *Annals of the New York Academy of Sciences* 1000: 7–31.
Flombaum, Jonathan I. and Santos, Laurie R. 2005. "Rhesus monkeys attribute perceptions to others." *Current Biology* 15: 447–452.
Genty, Emily, Breuer, Thomas, Hobaiter, Catherine and Byrne, Richard W. 2009. "Gestural communication of the gorilla (*Gorilla gorilla*): Repertoire, intentionality and possible origins." *Animal Cognition* 12: 527–546.
Hesler, Nana and Fischer, Julia. 2007. "Gestural communication in barbary macaques (Macaca sylvanus): An overview." In *The gestural communication of apes and monkeys*, Josep Call and Michael Tomasello (eds), 159–195. Mahwah: Lawrence Erlbaum Associates.
Hobaiter, Catherine and Byrne, Richard W. 2011. "The gestural repertoire of the wild chimpanzee." *Animal Cognition* 14: 745–767.
Kawai, M. 1965. "Newly acquired pre-cultural behavior of the natural troop of Japanese monkeys on Koshima island." *Primates* 6: 1–30.
Krützen, Michael, van Schaik, Carel and Whiten, Andrew. 2007. "The animal cultures debate: Response to Laland and Janik." *Trends in Ecology and Evolution* 22: 6.
Kummer, Hans and Goodall, Jane. 1985. "Conditions of innovative behaviour in primates." *Philosophical Transactions of the Royal Society of London B* 308: 203–214.
Laidre, Mark E. 2005. "Honest signaling of intentions: The importance of performance risk, receiver response, and prior knowledge of opponents in the threat displays of mandrills." *Behaviour* 142: 455–476.
Laidre, Mark E. 2006. "Manipulation without mind-reading: Information suppression and leakage during food discovery by mandrills (*Mandrillus sphinx*)." *Behaviour* 143: 365–392.
Laidre, Mark E. 2008. "Do captive mandrills invent new gestures?" *Animal Cognition* 11: 179–187.
Laidre, Mark E. 2009. "Informative breath: Olfactory cues sought during social foraging among old world monkeys (*Mandrillus sphinx, M. leucophaeus,* and *Papio anubis*)." *Journal of Comparative Psychology* 123: 34–44.
Laidre, Mark E. 2011. "Meaningful gesture in monkeys? Investigating whether mandrills create social culture." *PLoS ONE* 6: e14610.
Laidre, Mark E. and Yorzinski, Jessica L. 2005. "The silent bared-teeth face and the crest-raise of the mandrill (*Mandrillus sphinx*): A contextual analysis of signal function." *Ethology* 111: 143–157.

Laidre, Mark E. and Yorzinski, Jessica L. 2008. "Offspring protection by male mandrills, *Mandrillus sphinx*." *Primate Report* 76: 1–8.

Laland, Kevin N. and Galef, Bennett G. (eds). 2009. *The question of animal culture*. Cambridge: Harvard University Press.

Laland, Kevin N. and Hoppitt, William. 2003. "Do animals have culture?" *Evolutionary Anthropology* 12: 150–159.

Liebal, Katja, Müller, Cornelia and Pika, Simone (eds). 2007. *Gestural communication in nonhuman and human primates*. Amsterdam/Philadelphia: John Benjamins Publishing Company.

Lorenz, Konrad Z. 1952. *King Solomon's ring*. New York: Crowell.

Maestripieri, Dario. 1999. "Primate social organization, gestural repertoire size, and communication dynamics: A comparative study of macaques." In *The origins of language*, Barbara J. King (ed.), 55–77. Santa Fe: SAR Press.

Martin, Paul and Bateson, Patrick. 1993. *Measuring behaviour*. Cambridge: Cambridge University Press.

Paukner, Annika and Anderson, James R. 2006. "Video-induced yawning in stumptail macaques (*Macaca arctoides*)." *Biology Letters* 2: 36–38.

Paukner, Annika, Suomi, Stephen J., Visalberghi, Elisabetta and Ferrari, Pier F. 2009. "Capuchin monkeys display affiliation toward humans who imitate them." *Science* 325: 880–883.

Pika, Simone and Liebal, Katja (eds). 2012. *Developments in Primate Gesture Research*. Amsterdam/Philadelphia: John Benjamins Publishing Company.

Pika, Simone and Mitani, John. 2006. "Referential gestural communication in wild chimpanzees (*Pan troglodytes*)." *Current Biology* 16: R191–R192.

Pika, Simone, Liebal, Katja, Call, Josep and Tomasello, Michael. 2005. "The gestural communication of apes." *Gesture* 5: 41–56.

Pollick, Amy S. and de Waal, Frans B. M. 2007. "Ape gestures and language evolution." *Proceedings of the National Academy of Sciences of the United States of America* 104: 8184–8189.

Scott, Nicole M. and Pika, Simone. 2012. "A Call for Conformity: Gesture Studies in Human and Non-Human Primates". In *Developments in Primate Gesture Research*, Simone Pika and Katja Liebal (eds), 145–161. Amsterdam/Philadelphia: John Benjamins Publishing Company.

Smuts, Barbara. 2002. "Gestural communication in olive baboons and domestic dogs." In *The cognitive animal*, Marc Bekoff, Colin Allen and Gordon M. Burghardt (eds), 301–306. Cambridge: MIT Press.

Struhsaker, Thomas T. 1967. "Behavior of vervet monkeys and other cercopithecines." *Science* 156: 1197–1203.

Tomasello, Michael and Call, Josep. 1997. *Primate Cognition*. New York: Oxford University Press.

Tomasello, Michael, Call, Josep, Warren, Jennifer, Frost, G. Thomas, Carpenter, Malinda and Nagell, Katherine. 1997. "The ontogeny of chimpanzee gestural signals: a comparison across groups and generations." *Evolution of Communication* 1: 223–259.

Whiten, Andrew and Mesoudi, Alex. 2008. "Establishing an experimental science of culture: animal social diffusion experiments." *Philosophical Transactions of the Royal Society of London B* 363: 3477–3488.

Whiten, A., Goodall, J., McGrew, W. C., Nishida, T., Reynolds, V., Sugiyama, Y., Tutin, C. E. G. Wrangham, R. W. and Boesch, C. 1999. "Cultures in chimpanzees." *Nature* 399: 682–685.

Gesture use in consortship

Wild chimpanzees' use of gesture for an 'evolutionarily urgent' purpose

Catherine Hobaiter & Richard W. Byrne
Centre for Social Learning and Cognitive Evolution and Scottish Primate Research Group, University of St Andrews, School of Psychology, St Andrews, KY16 9JP, Scotland

We describe the gestural communication of wild chimpanzees in the evolutionarily urgent context of consortship. Consortship represented the dominant context for the use of gestural communication by adult males in the Sonso community. Gesturing provided consorting males with the opportunity to communicate their intentions to the female, while avoiding the risk of also advertising these to other community males. The extensive use of gesturing by adult males differs from the low frequency reported in captivity, and emphasises the importance of studying behavior in a natural population.

1. Introduction

Play has repeatedly been found the most prolific context for the use of gestural communication by great apes in captivity, where most study of great ape gesture has taken place (Liebal et al. 2004; Tomasello et al. 1994; Genty et al. 2009; Pika et al. 2003; Pika et al. 2005). In consequence, it has been suggested that gestural communication is generally used for "less evolutionary urgent functions" (Tomasello & Call 2007: 5). But it is not clear which, if any, contexts experienced by captive apes would require communication about evolutionarily urgent functions; in other words, those that have been subject to strong selection pressures. In contrast, wild chimpanzees patrolling their territorial boundaries, hunting, or initiating consortship behavior, run very real risks: up to and including their own death. Thus, they might be expected to employ communicative strategies that minimize these risks. In a secondary rainforest where visual lines of sight can be restricted, vocalization represents an effective means of communication. However, with all vocalizations there is a risk that the call may be overheard and the information employed by unintended recipients 'eavesdropping' – particularly where the individual calling can also be identified (Peake et al. 2002, Mennill et al. 2002). Unlike vocalizations, silent and contact gestures allow the signaller to communicate their intention without the risk of that message being 'overheard' by other parties. Unfortunately, its inherently secret nature means that to date there has been very little empirical analysis of consortship behavior. Here, we take advantage of a recent cluster of observations in the Sonso community, Budongo Forest Reserve, Uganda, to discuss the nature of consortships and the role of gestural communication within them in more detail.

1.1 Chimpanzee sexual strategies

Early work in this field emphasised male sexual strategy (Allen 1981; Hasegawa & Hiraiwa-Hasegawa 1983; Tutin 1979; Tutin & McGrew 1973); thus, sexual behavior was defined in terms of male-male competition as either opportunistic (non-competitive mating, with free access to all males), or restrictive (access to the female being monopolised by a single male). Within the category of restrictive mating, we can discriminate two patterns of behavior: *possessiveness*: where sexual access to a female beings monopolized by a single male while remaining within the group (also known as *mate-guarding*); and *consortship*: where a single male monopolizes sexual access by escorting a female away from the group (Tutin 1979). At Gombe, Tutin found that consortships were associated with a high probability of reproductive success (Tutin 1979). A genetic analysis of the Taï community, however, suggests that this may vary between males (Boesch & Boesch-Achermann 2000).

More recently, work on sexual strategy has emphasised the role of female choice in determining paternity (Pieta 2008; Stumpf & Boesch 2006; Emery Thompson et al. 2008; Boesch 2009; Stumpf & Boesch 2010). This is particularly true in the case of consortship, where the consorting male must avoid detection by other group males: even a brief scream from the female may bring other males to investigate, particularly if she is known to be in oestrus. Aggressive coercion by the male is often observed in the initial stages of consortship (Goodall 1986), apparently to overcome reluctance on the part of the female. This has suggested that promiscuity represents a more favourable strategy for female chimpanzees (Muller et al. 2007). However, a recent study showed that aggressive male coercion did not in fact act to decrease female resistance (Stumpf & Boesch 2010). In addition, Nishida (1997) described females at Mahale responding with "blunt refusal of male courtship." Co-operation on the part of the female may then be critical to the success of a consortship. Tutin (1979) observed that males who frequently engaged in grooming and sharing food with oestrus females while they were with the group were more likely later to be successful in leading females away from the group on consortship. Furthermore, Goodall (1986) described the use of grooming in consortship to reduce the anxiety of a reluctant female, making her easier to lead away.

Consorting males must invest time and energy in removing the female from the group. Lower-ranking individuals may have to initiate a consortship several days before a female reaches peak fertility, as she is then maximally capable of conception and likely to be surrounded by other, more dominant would-be suitors. In addition to the time and energy invested, consortships are also associated with increased risk. A consorting male risks aggressive attack from males within the community, should they discover his attempt to remove the female or when he tries to rejoin the group after the absence while on consortship (Riss unpublished cited in: Tutin 1979). Yet, in attempting to avoid detection by their own group, consorting pairs are more likely to occupy peripheral areas of the community territory, increasing their chances of encountering neighbouring groups. Both male and female risk attack from individuals of neighbouring communities. Tutin (1979) for example cites intercommunity encounters as the highest source of risk for individuals already on consortship. For the male this may present a direct risk to his life, as intercommunity encounters can result in lethal aggression (Goodall 1986). Females, particularly those in oestrus, are less likely to be killed, but may be aggressively herded into the neighbours' community (Boesch 2009). There they risk attack by the community females (Townsend et al. 2007), and any dependent offspring travelling with them may be killed (Suzuki 1971; Reynolds 2005; Townsend et al. 2007).

1.2 The potential role of gestural communication when on consortship

The decision to initiate a consortship may evidently depend on a number of factors including male rank, female co-operation, and risks associated with intercommunity encounters. However, once the decision has been made, in all cases there is significant pressure on a consorting male to communicate his initial intention to the female in a discreet manner, and for both the male and female to avoid detection once in consortship. For these reasons, use of gestural rather than visual communication may represent an effective strategy. Chimpanzees not only regularly use gesture to communicate their desires, but they intentionally alter the modality of their gestures with respect to other individuals' states of attention (Tomasello & Call 2007; Genty et al. 2009; Liebal et al. 2004; Pika et al. 2003; Hobaiter & Byrne 2011a). We hypothesised that, because of the need to limit the communication to a specific recipient, gestural communication – particularly silent and contact based gestures – would be favoured in the consortship context. Due to a recent peak in consortship behavior within the Sonso community, we are able to report that high levels of gestural communication did indeed occur during these consortships, and we describe the nature of the interactions, including gestures and vocalizations.

2. Method

2.1 Procedure

Observations of consortship behavior were recorded on an ad-hoc basis during systematic data collection for a project on chimpanzee gestural communication among the wild Sonso chimpanzee community in the Budongo forest, Uganda at the Budongo Conservation Field Station (BCFS) (Hobaiter & Byrne 2011a). Observations were made during 18-months of observation, split into three periods between October 2007 and August 2009. All examples of consortship behavior (as defined below) were recorded on miniDV using a Sony Handycam (DCR–HC-55).

2.2 Long-term data collection

In addition to direct observations, we interrogated the six highly-experienced, chimpanzee field-assistants (two of whom have worked with the Sonso community for 20-years), in order to establish a long-term record of consortship frequency. We also consulted the BCFS events book, kept on site for the purpose of collating unusual or rare observations. Field-assistants record the frequency and duration of aggressive behavior ad libitum onto handheld Workabout-Pro computers; these are collated in the projects long-term records (Zuberbühler & Reynolds 2005).

2.3 Defining sexual behavior

We follow Tutin (1979), in defining *consortship* as: "where a single adult male escorts a female away from the group and maintains exclusive copulatory access to her" (Tutin 1979: 30). We define a consortship as successful when the female was isolated from the group and the pair remained absent for a minimum of 48-hours.

2.4 Defining gestures

We define gestures as discrete, mechanically-ineffective physical movements of the body observed during periods of intentional communication. Thus, each case of gesture had to be accompanied by an indication of intentional use. We considered gestures accompanied by one or more of the following parameters to involve intentional communication:

Audience-checking: the signaler shows signs of being aware of the potential recipients and their state of attention, e.g. turning to look at the recipient before gesturing.

Response-waiting: After gesturing the signaler pauses for >1 sec and maintains some visual contact.

Persistence: the production of further gestures after response-waiting.

Where a string of gestures, separated by <1sec, was followed by response-waiting, we assigned the intentional aspect to each gesture within the string.

2.5 Structure of the gestural communication

As per Hobaiter & Byrne 2011b, we define the following structures within gesturing:

Single gesture: a single gesture followed by a pause of >1sec of response-waiting.
Rapid sequence: multiple gestures without intermittent pauses of >1sec.
Bout: multiple single gestures and/or rapid sequences produced in succession with intermittent pauses of >1sec and/or non-gestural behavioral responses from the recipient.

2.6 Function of the gestural communication

Function was defined by the behavioral response that led to the end of the communication attempt (as per. Genty et al. 2009). Function was measured at the level of the bout: we considered all single gestures and rapid sequences within a bout to be produced for the same function.

2.7 Success of the gestural communication

We considered persistence in communication to imply the failure of earlier gestures. Where a response appeared to satisfy the gestural communication, the single gesture or rapid sequence immediately preceding it was considered to be *successful*. Where the recipient produced a behavioral response that did not satisfy the signaler, but was congruent with a subsequent behavior that did, we considered the gesture or rapid sequence to be *partially successful*. For example: a signaler gestures in a rapid sequence towards an inattentive recipient, the recipient looks round and moves towards the signaler but stops short of reaching him, the signaler gestures with another rapid sequence and the recipient then moves to play with the signaler, the signaler then stops gesturing. Both rapid sequences in the bout would be considered to have the function of requesting play, the second sequence would be considered completely successful, the first sequence would be considered partially successful.

2.8 Gesture modality

We categorized gestures according to their potential mode of reception as signals, as *silent*, *audible* or *contact*. In the dense secondary rainforest many movements may result in a sound being produced. However, we treated gestures as audible only when they were made audible by their intrinsic features, i.e. that they would be audible in every case, irrespective of where or when they were produced.

2.9 Long and short-distance audible gestures

In rainforests, the complicated acoustic environment leads to increased degradation and attenuation of acoustic signals (Wiley 1991). Mitani and colleagues (1999) found that the pant-hoot calls of different chimpanzee populations varied in a manner that maximized signal transmission with variation in the habitat acoustics. In dense, secondary rainforest such as that found at Budongo, the degradation of acoustic signals would be particularly high. Studies of chimpanzee vocal behavior typically distinguish between short and long-distance chimpanzee vocal behavior (e.g. pant-grunt vs. pant-hoot, see: Van Lawick-Goodall 1972; Crockford & Boesch 2005), and we suggest that it is possible to distinguish audible gestures in the same way.

Although clearly audible, *Object-move* and *Object-shake* gestures appear to be limited in terms of their audible range. For example: when the highly experienced head field-assistant was trying to locate a consorting male whom we observed to repeatedly produce *Object-shake* gestures, he failed to do so until less than 100m away, despite awareness of the approximate location. In addition to the short range over which they can be heard, the audible component of these gestures comes from the rattling of leaves and foliage, something that can be caused by other large forest mammals such as bush-pigs (or field-researchers); and as such they are not immediately acoustically identifiable as chimpanzee gestural communication.

In contrast, certain gestures are audible over much greater distances, and are purely associated with chimpanzee communication. These are the *Drum-object* or *Stomp-on-object* gestures. When the object in question is one of the large tree buttresses regularly found throughout the forest, and regularly employed by the chimpanzees for drumming and stomping actions, the gestures produce a distinctive deep boom that is audible to humans over 500m away. In many cases, individual idiosyncrasies allow us to identify not only the location but also the identity of the drumming chimpanzee, – a highly effective long-distance signal (Clark Arcadi et al. 1998). Several observations of solitary male chimpanzees repeatedly drumming and then waiting until there is a response from a party of chimpanzees before moving directly to them, suggests that the chimpanzees themselves are aware of the long-distance quality of these communications. Furthermore observations of the immediate change in behavior, when the drum of an individual from a neighbouring group is heard, strongly suggest that chimpanzees are also capable of distinguishing individuals in this manner, – observation supported by similar reports from chimpanzees in the Taï forest (Boesch & Boesch-Achermann 2000).

2.10 Specific analyses

Data were converted to means for each individual, to remove any effect of pseudo-replication from the use of focal behavior sampling. Only individual means calculated from five or more separate cases were included in any analyses. Analyses were carried out in SPSS v11, with $\alpha = 0.05$ required for significance. Means are given ±Standard Deviation, throughout. Data were all examined for appropriateness for parametric statistics and where necessary we applied transformations applied and re-tested the data. Where transformations were applied the results are clearly labelled; where no appropriate transformations were possible non-parametric alternatives were used. Statistical tests are two-tailed.

3. Results

3.1 Consortship behavior in Sonso chimpanzees

Consortship behaviour was rarely observed, with only ten events reported in the past ten years. During 266 days of observation between October 2007 and August 2009, we observed four cases of consortship behavior in the Sonso community involving two males and three females, and were able to record over two hours of video footage (total 2h 24m 37s: includes 17m30s kindly donated by other researchers).

1. 05.01.2008 Duane and Lola (45m 50s video) unsuccessful
2. 20.01.2008 Duane and Zimba (53m 15s) successful
3. 03.02.2008 Duane and Zimba (28m 2s) successful
4. 04.10.2008 Nick and Nambi (17m 30s) successful

3.2 Gestural communication in the consortship context

In the 18-month study of gesture in the Sonso community, consortships were recorded on only four of the 266 days of observation (1.50%) but accounted for 412 of the 4397 gestures recorded (9.39%). Critically, gestures from consortships accounted for 62.18% of all adult male gesture use recorded during the study (393/632 gestures), with males producing almost all of the gestures used in this context (n = 412, males: 393, females: 19). Gesturing was recorded both when the male and female were still within the group, and also once the pair had moved away from the group, but were still within the core area of the Sonso community.

3.3 The consortship repertoire of gestures

The complete Sonso gestural repertoire consists of 66 types of gesture, used flexibly across ten different contexts (Hobaiter & Byrne 2011a). Twenty-one of these gesture types were recorded during *consortship*, 17 from males and only four from females (predominantly the *Present-sexual* gesture, 16 of the 19 cases of female gesture). The most frequently used gestures were the object related gestures: *Object-shake* (222 cases), and *Object-move* (41 cases), which together accounted for over 60% of all gestural communication in this context. We observed no consortship-specific gestures. However, the rare *Rump-rub* gestures were predominantly produced within the consortship context (26/29 cases).

In *Rump-rub* the male signaler backs up to the recipient and pushes his rump against him (usually their genitals or torso); this movement is accompanied by a small but rapid, vertical up-and-down rubbing motion. *Rump-rubs* were often accompanied by a soft-pant vocalization. Twenty-six cases were recorded during consort behavior. In other contexts (two Agonistic, one Unknown), the gesture was used by a less dominant male to a more dominant male when apparently seeking affiliation or reassurance. However, in the consortship context a dominant male directed the gesture to a lower-ranking female.

3.4 Gesture as discreet communication?

Perhaps surprisingly, audible gestures were extremely prevalent in consortship communications (334/412, 81.1%); even silent and contact gestures were accompanied by audible behaviors in a third of all observed cases (26/78). Within male gestural communication, 85.0% of gestures were audible gesture types (334/393, *Duane*: 321/379; *Nick*: 13/14), and 91.6% were either audible or accompanied by other audible behavior (360/393, *Duane*: 347/379, *Nick*: 13/14). This actually represented an increase in the proportional use of audible gestures over use in other contexts, by both males (*Duane* consortship audible = 321/379, non-consortship audible = 7/17, Chi-square $\chi^2 = 21.67$, df = 1, $p < 0.0001$. *Nick* consortship audible = 13/14, non-consortship audible n = 59/106; Chi-square $\chi^2 = 4.60$, df = 1, $p = 0.0319$).

The prevalence of audible gestures was due to the prolific use of the *Object-shake* and *Object-move* gestures described above; but critically these are all short-distance audible gestures. In other contexts, short-distance audible gestures (*Object-shake* and *Object-move*) were used in the same rapid sequence as long-distance audible gestures (*Drum-object* or *Multiple stomp-on-object*) in a mean 6.7% of cases (19/285). However, despite the prevalence of the short-distance *Object-shake* and *Object-move* gestures in consortship communications, there were no cases of long-distance audible gestures during consortships (0/211) (n = 496, Chi-square: $\chi^2 = 14.06$, df = 1, $p = 0.0002$).

3.5 The function of gestural communication in consortship

Consortship communication included 127 separate bouts of gesturing, 61 of which were successful and could therefore be used to define function. Unsurprisingly, given the nature of consortship behavior, the overwhelming majority of the gestural communications produced by both males had the apparent function that the female should 'follow' him (*Duane*: 48/52 bouts, 92.3%; *Nick*: 5/6 bouts, 83.3%). Perhaps more surprisingly, only a very low number of bouts (total two: *Lola* one; *Duane* one) were used for the function of acquiring 'sexual attention' (this function includes both inspection and copulation). *Leaf-clipping*, a gesture that was closely associated with the function of acquiring 'sexual attention' outside of consortship (31 of 40 recorded cases), was never observed during consortship communication. Other recorded functions included: Affiliation, Direct attention, Move closer, Position and Stop behavior (all single cases).

The function 'follow' was very rarely recorded outside of the Consortship context, and never from adult males. During the study it was recorded in only eight other bouts, all in 'travelling'; and almost all were mother to offspring communications (6/8; also one case between two brothers, one case between two sub-adults).

3.6 Response-waiting in gestural communication on consortship

Response-waiting is an indication of intentional communication and as such was one of several criteria for intentionality within this analysis. However, response-waiting was not the only indication used, so its distribution might still vary within the overall data set. Indeed, both adult males employed response-waiting significantly more often following consortship communications compared with other contexts (*Duane* response-waiting: consortship = 345/379, mean frequency = 88.7%; other contexts: 5/17, mean frequency = 29.4%, Chi-square $\chi^2 = 7.66$, $df = 1$, $p = 0.0057$; *Nick* response-waiting: consortship = 14/14, mean frequency = 100.0%; other contexts = 40/89, mean frequency = 44.9%, Chi-square $\chi^2 = 14.70$, $df = 1$, $p = 0.0001$).

3.7 Success of male communications in consortships

Figure 1 illustrates the variation in frequency of success of gestural communications *from* males, within and outside of the consortship context, alongside the variation in frequency with which females provided a successful response *to* gestural communications, within and outside of the consortship context. In both cases, the level of any kind of success seems to be lower within consortships than at other times. This is particularly so when indexed by the rates of completely successful communication.

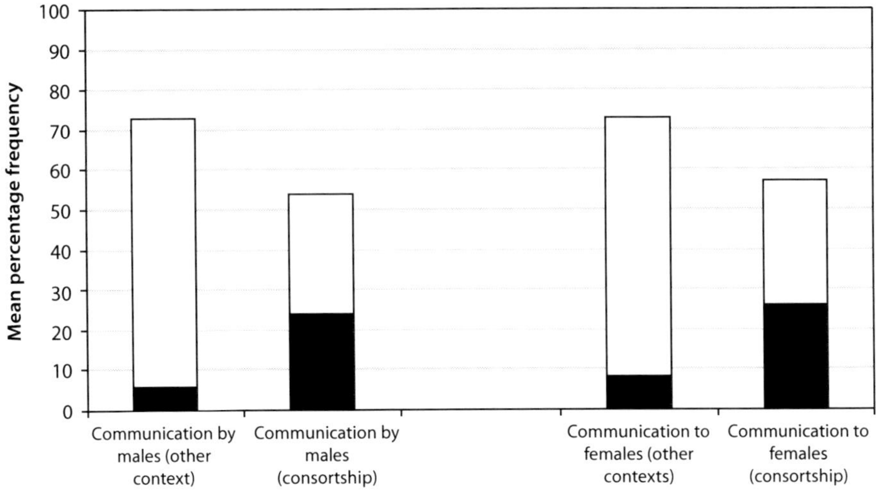

Figure 1. Variation in the frequency of success in communications *given* by males and *received* by females. The percentage of gestural communications produced by adult males that receive a successful or partially successful response is compared with the number of communications to which an adult female responds successfully or partially successfully in consortship behavior and other contexts. Black bars represent partially successful communications; white bars represent completely successful communications. Note that categorizations 'by males' and 'to females' are not identical because of the use of individual means to control for pseudoreplication.

3.7.1 *Full success*

Duane experienced significantly lower success in consortship communication (successful gestural communications: consortship n = 229, mean frequency = 25.3%; other contexts n = 16, mean frequency = 75.0%. Fisher's exact test $p < 0.0001$). There was no significant variation in the success of the alpha male *Nick* (successful gestural communications: consortship n = 14, mean frequency = 42.9%; other contexts n = 68, mean frequency = 58.8%. Fisher's exact test p = 0.377.)

3.7.2 *Partial success*

Again the alpha male *Nick* experienced no variation in partial success between consortship and other communications. However, *Duane* experienced a significant increase in partially successful communication. (*Duane* partially successful gestural communications: consortship n = 229, mean frequency = 33.6%; other contexts n = 16, mean frequency = 0.0%. Fisher's exact test p = 0.004. *Nick* partially successful gestural communications: consortship n = 14, mean frequency = 14.3%; other contexts n = 68, mean frequency = 13.2%. Fisher's exact test p = 1.000).

3.8 Female responsiveness to male gestural communication on consortship

The variation in rate of success and partial success experienced by the two consorting males may be due to a difference between the males (e.g. rank) or a difference between the females with whom they attempted consortship. Table 1 describes the variation in female responsiveness to gestural communication while on consortship when compared to that in other contexts.

The only significant change in behavior was recorded from the female *Zimba*, with whom *Duane* consorted twice. She produced complete responses significantly less often when consorting, although her level of partial responses was then higher than usual, suggesting that her responses were often not outright refusals.

Table 1. Variation in individual female responsiveness. The percentage of communications to which each individual female responded in a successful or partially successful fashion, when on consortship and in other contexts. Raw data given in brackets.

Female	Consortship		Other contexts		Fisher's exact test
	n	Mean frequency %	n	Mean frequency %	p
Successful response					
Zimba	159	33.3 (53/159)	39	51.3 (20/39)	0.039*
Lola	70	7.1 (5/70)	11	27.3 (3/11)	0.072
Nambi	13	38.5 (5/13)	66	68.2 (45/66)	0.060
Partially successful response					
Zimba	159	38.4 (61/159)	39	5.1 (2/39)	<0.0001*
Lola	70	22.9 (16/70)	11	9.1 (1/11)	0.440
Nambi	13	15.4 (2/13)	66	6.1 (4/66)	0.255

Note: * represents a significant variation between behavior in consortships and other contexts.

Lola produced very low levels of successful responses when in consortship (less than a third of either of the other two females) but this was not significantly lower than her level of response outside of consortships. In addition, her level of partial responses was not increased, as observed in *Zimba*. *Nambi* also made no significant change in her behavior, although her level of successful response was much higher than that of *Lola*, thereby being equivalent to that of *Zimba*.

3.9 Persistence in gestural communication on consortship

Sonso chimpanzees persist following the failure ($n = 41$, mean frequency $= 48.02\% \pm 20.43$) and, in particular, the partial failure ($n = 23$, mean frequency $= 71.31\% \pm 15.97$) of a gestural interaction (Independent t-test: $t = 4.76$, $df = 62$, $p < 0.0001$). When compared with communication in other contexts, persistence following total failure was significantly higher in consortship communication by *Duane*, and approached a significant increase in *Nick* (see Table 2). Persistence following partial success in consortship behavior was high from both males, but the available data were limited and there was no significant variation between this and other contexts.

Table 2. Variation in individual male persistence. The percentage frequency with which each individual male persisted in further communication following the partial or total failure of the earlier communication when on consortship or in other contexts is compared using Fisher's exact tests.

Persistence following:	*Duane*			*Nick*		
	Consortship (%)	**Other contexts %**	**Fisher's exact p**	**Consortship %**	**Other contexts %**	**Fisher's exact p**
Total failure	120/162 (74.1)	3/5 (60.0)	<0.0001	6/6 (100.0)	10/20 (50.0)	0.0532
Partial failure	98/114 (86.0)	*	–	2/2 (100.0)	5/11 (45.5)	1.00

Note: * represents no cases observed in the category.

3.10 Vocalizations during consortship

3.10.1 *Male vocalization*

Neither male produced any loud vocalization while on consortship. *Duane* was observed to produce soft-pants in connection with the *Rump-rub* gesture, *Nick* was not observed to vocalize while on consortship.

3.10.2 *Female vocalization*

Nambi was not observed to vocalize loudly when in consortship with *Nick*. *Lola* and *Zimba* both produced loud vocalizations, including eleven bouts of screaming (*Zimba* = 6, *Lola* = 5), in the early stages of consortship; all of these were followed by (9) or produced during (2) an aggressive attack from *Duane*. On the first consortship between *Duane* and *Zimba*, screaming by *Zimba* resulted in the consorting pair being discovered by a group of males. On this occasion, all the newly arrived males were subordinate to the consorting male *Duane* and after a brief period of rest he escorted *Zimba* away again. The screaming from *Lola* is likely to have contributed to the eventual location of the pair by a party containing the alpha male, which resulted in the immediate termination of the consortship as *Duane* was chased away. *Zimba* was frequently observed to produce a soft-bark immediately following a bout of gestural communication from *Duane*; she would then follow on shortly after. This appeared to satisfy *Duane*, who would only resume gesturing if she continued to then make no further move towards him.

3.11 The coercion of females on consortship

During the year 2008, a total of 178 physically aggressive attacks were recorded in the Sonso community. Sixty-three of these attacks were classed as high-intensity attacks due to repeated physical contact (hitting, kicking, biting etc.) and/or a resultant physical injury. Twenty-five of the high-intensity attacks were directed to females during consortships (39.7% of high-intensity attacks on 1.1% of observation days).

The majority of the high-intensity physical attacks (17/25) followed a failure by the female to respond to the male's gestural communication, predominantly a request that the female follows him. Three high-intensity attacks followed an attempt by the female to communicate vocally with other group members, and a further three followed attempts by the female to move away from the male.

Duane groomed the female on all three consortships but to a varying degree. In his consortship with *Lola* he aggressively coerced her from the start, and grooming behavior was negligible (two bouts both <10 seconds). During the first consortship with *Zimba* he was discovered in the morning grooming her in her sleeping nest. He groomed her briefly once when she climbed down, and then following a period of more aggressive coercion he started grooming her for longer periods (>5 min). On the second consortship, the beginning was again marked by a brief period of aggressive coercion followed by long bouts of grooming (>10 min) once they were away from the group. *Nick* was not observed to groom *Nambi*.

4. Discussion

Consortships represent an understudied area of chimpanzee behavior. Irregular, infrequent, and inherently secret, they are particularly difficult to record. Nevertheless, they provide crucial insight into an unusual form of social relationship in chimpanzee behavior: an isolated male-female pair. Consortships are rare in the Sonso community (on average ~1 per year: slightly higher than the rate reported at Mahale (Hasegawa & Hiraiwa-Hasegawa 1983), but much lower than Taï (Boesch & Boesch-Achermann 2000) or Gombe (Tutin 1979)). However, despite the low frequency of consortship events, gestural communication was used prolifically within them, and indeed consortships represented the dominant context for gestural use by adult male chimpanzees. This enabled us to compare the gestural communication of individuals within the context of consortship with that made in other contexts. Male chimpanzees produced almost all of the gestural communication within consortships; they used a range of 17 gesture types, predominantly to request that the female follow them away from the group. In order to maintain exclusive access to female at her point of peak fertility, it is necessary to remove her before she reaches this stage. It is thus logical that the immediate function of the male's communication is to take the female away with him, rather than to facilitate engagement in sexual behavior. The apparent lack of interest in immediate sexual access is supported by the absence of *leaf-clipping*, which is commonly used by mature Sonso chimpanzees to request sexual attention. Almost all gestures with the function of obtaining sexual attention were produced by the females, and accounted for most female gestural communication in consortships.

Perhaps surprisingly, considering the pressure to avoid detection by other chimpanzees (either by other Sonso males, or other communities), the majority of gestures used in consortship were audible. Given the loss of investment and the physical risk, should other individuals become aware of the consorting male's intentions, the use of audible gestures seemed initially counterintuitive. In attempting to understand this, we developed the post-hoc hypothesis that, as with vocalizations, there may be different levels of 'audibility' within gestural communication. Audible gesturing while in consortship was restricted to relatively quiet audible gestures such as *Object-shakes*. These gestures were not only limited in the range that their sound would travel, but the audible component was an extremely discreet one: rustling foliage. Furthermore, unlike a vocalization or high-amplitude gesture such as a *buttress drum*, these short-distance, low-amplitude gestures do not reveal individual identity; reducing the risk to the male signaling should another individual, particularly another more dominant male, 'overhear' the message. In contrast, in consortships there was no use of the long-distance drumming gestures that are immediately identifiable as chimpanzee communication, and from which it is also possible to discern individual identity (Arcadi et al. 1998). Thus, the gesturing still appeared to provide a discreet means to communicate in the dense secondary-forest environment. This use of short-distance audible communication in gesturing mirrors the pattern of vocalizations produced during consortship behavior. Males were only observed to produce soft-pant vocalizations and we found that, as McGinnis (1973) reported, loud vocalizations on the part of the female resulted in the threat or use of physical violence by the male. This supports the idea that consorting male chimpanzees are highly motivated to employ discrete methods of communication.

In comparison to contexts such as play or grooming, sexual behavior – in particular risky sexual behavior such as consortship – represents an evolutionarily more urgent function. Although the scarcity of the behavior limited the number of events and individuals available for analysis, the high levels of gesturing allowed us to compare the gestural behavior within consortships with that of the same individuals in other contexts. Given the extremely large potential payoff of a successful consortship, consortship communication may be considered to have a high evolutionary value for the male. This increased value is reflected in increased response-waiting, and (to an extent) increased persistence following failed communications.

The motivation to succeed in a consortship, as evidenced by unusually high levels of response-waiting when gesturing, can also be judged by the level of aggressive coercion employed by consorting males. While brief fights are not unusual in chimpanzee behavior, prolonged bouts of severe aggression are rare, particularly between males and females. Extensive grooming of the female did occur on two successful consortships, but only after the female started to follow freely and the pair had moved away from the core Sonso territory. Thus, grooming appeared to function less as a form of coercion and more as a means to re-establish the social bond between the pair following the earlier aggression. This delay in grooming makes the repeated use of the rare *Rump-rub* gesture at the earlier, more aggressive, stage in the consortship particularly interesting. *Rump-rub* was documented at Gombe as *rump turning* (Goodall 1968) and in a captive group by Tutin and McGrew (1973) as *bump rump*. At Sonso the gesture is usually used outside of consortship, by a lower-ranking individual towards a more dominant one, in requests for affiliation or reassurance. Given the use of aggressive coercion towards a still-reluctant female, it seems puzzling that the dominant male would at this stage choose to employ a gesture whose physical form (a low crouch with the genitals exposed) is classically submissive, and that is associated with a submissive role in other contexts. However, the soft-pant vocalization that accompanied the gesture may give some clue as to his intention. This vocalization is normally given to a trusted group member in contexts such as food excitement or the arrival of a friendly other. The apparent submission implied by the use of the *Rump-rub* gestures by the male may represent an attempt to reassure the female, while continuing to gesture and communicate his desire that she should follow him. Goodall suggested that the use of grooming by a consorting male might relax the female and give "proof of his fundamentally friendly intent", making her easier to lead away (Goodall 1986: 402). However, grooming requires that both individuals are stationary for several minutes or more. While this may be appropriate once they have moved away from the main group, it could be extremely costly for the male to sit down while still 'under-the-nose' of the other community males, particularly if the female is still not entirely co-operative. Thus the *rump-rub* gesture may function as a 'quick-and-dirty' way of expressing the consorting male's essentially friendly intent: encouraging the female in the initial stages of consortship as she is being coerced into leaving the main group; but without the need for time-consuming bouts of grooming while the pair is still in the vicinity of other community males and risk discovery.

While on consortship, levels of female responsiveness to male gestural communication were generally low. However, one female, *Zimba*, although she infrequently responded with the desired 'follow', did, at least, frequently produce behavior that was congruent with following. In practice, this meant that although she did not follow the male she would turn to attend to his communication, and approach a little way towards him. While on consortship *Zimba* was accompanied by her two young sons who would occasionally wander behind, so her partial responses may have represented an attempt on her part to acknowledge *Duane's* communication while at the same time giving her sons time to catch up. In doing so *Zimba* avoided the complete refusals that were associated with aggressive coercion such as high-intensity physical attacks.

The extensive gesturing by adult males in consortship differs dramatically from the low frequency of adult male gesturing reported in captivity, and represents the first description of gesture use in an 'evolutionary urgent' context. Gestural communication offers male chimpanzees the opportunity to communicate their intentions to the female they wish to engage in consortship, while minimising the risk of also advertising these intentions to other community males. This finding highlights the importance of studying behavior in a natural population where the full range of environmental and social contexts is available.

Acknowledgements

We thank all BCFS staff, especially Amati Stephen, the project's founder Vernon Reynolds, and its current scientific director Klaus Zuberbühler. We thank Marion Laporte for her kind donation of additional video footage. We acknowledge the help of Dr Cathy Crockford and two anonymous referees, who provided valuable suggestions on an earlier manuscript. We thank the Uganda National Council for Science and Technology, and the Uganda Wildlife Authority. Fieldwork of CH was generously supported by grants from the Wenner-Gren Foundation and the Russell Trust.

References

Allen, Mel. 1981. "Individual copulatory preference and the "strange female effect" in a captive group-living male chimpanzee (*Pan troglodytes*)." *Primates* 22 (2): 221–236.

Boesch, Christophe. 2009. *The Real Chimpanzee: Sex strategies in the forest.* Cambridge, UK: Cambridge University Press.

Boesch, Christophe and Boesch-Achermann, Hedwige. 2000. *The Chimpanzees of the Taï Forest: Behavioural ecology and evolution.* Oxford, UK: Oxford University Press.

Clark Arcadi, Adam, Robert, Daniel and Boesch, Christophe. 1998. "Buttress drumming by wild chimpanzees: temporal patterning, phase integration into loud calls, and preliminary evidence for individual distinctiveness." *Primates* 39: 503–516.

Crockford, Catherine and Boesch, Christophe. 2005. "Call combinations in wild chimpanzees." *Behaviour* 142: 397–421.

Emery Thompson, Melissa, Stumpf, Rebecca M. and Pusey, Anne E. 2008. "Female reproductive stategies and competition in apes: An introduction." *International Journal of Primatology* 29: 815–821.

Genty, Emilie, Breuer, Thomas, Hobaiter, Catherine and Byrne, Richard W. 2009. "Gestural communication of the gorilla (*Gorilla gorilla*): Repertoire, intentionality and possible origins." *Animal Cognition* 12 (3): 527–546.

Goodall, Jane. 1986. *The Chimpanzees of Gombe*. Cambridge, MA: Harvard University Press.

Hasegawa, Toshikazu and Hiraiwa-Hasegawa, Mariko. 1983. "Opportunistic and restrictive matings among wild chimpanzees in the Mahale Mountains, Tanzania." *Journal of Ethology* 1: 75–85.

Hobaiter, Catherine and Byrne, Richard W. 2011a. "The gestural repertoire of the wild chimpanzee." *Animal Cognition* 14: 745–767.

Hobaiter, Catherine and Byrne, Richard W. 2011b. "Serial gesturing by wild chimpanzees: its nature and function for communication." *Animal Cognition* 14: 827–838.

Liebal, Katja, Call, Josep and Tomasello, Michael. 2004. "Use of gesture sequences in chimpanzees." *American Journal of Primatology* 64 (4): 377–396.

McGinnis, Patrick R. 1973. "Patterns of sexual behaviour in a community of free-living chimpanzees." Ph.D. dissertation. Cambridge University, Cambridge.

Mennill, Daniel J., Ratcliffe, Laurene M. and Boag, Peter T. 2002. "Female eavesdropping on male song contests in songbirds." *Science* 296: 873.

Mitani, John C., Hunley, Keith L. and Murdoch, M. E. 1999. "Geographic variation in the calls of wild chimpanzees: A re-assessment." *American Journal of Primatology* 47: 133–152.

Muller, Martin N., Kahlenberg, Sonya M., Emery Thompson, Melissa and Wrangham, Richard. 2007. "Male coercion and the costs of promiscuous mating for female chimpanzees." *Proceeding of the Royal Society B* 204: 1009–1014.

Nishisa, Toshisada. 1997. "Sexual behavior of adult male chimpanzees of the Mahale Mountains National Park, Tanzania." *Primates* 38 (4): 379–398.

Peake, Thomas M., McGregor, Peter K. and Dabelsteen, Torben. 2002. "Do great tits assess rivals by combining direct experience with information gathered by eavesdropping?" *Proceedings of the Royal Society of London B* 269: 1925–1929.

Pieta, Katharin. 2008. "Female mate preferences among *Pan troglodytes schweinfurthii* of Kanyawara, Kibale National Park, Uganda." *International Journal of Primatology* 29: 845–864.

Pika, Simone, Liebal, Katja and Tomasello, Michael. 2003. "Gestural communication in Young Gorillas (*Gorilla gorilla*): Gestural Repertoire, Learning and Use." *American Journal of Primatology* 60: 95–111.

Pika, Simone, Liebal, Katja and Tomasello, Michael. 2005. "Gestural communication in subadult bonobos (*Pan paniscus*): Gestural repertoire and use." *American Journal of Primatology* 65 (1): 39–61.

Reynolds, Vernon. 2005. *The chimpanzees of the Budongo Forest*. Oxford, UK. Oxford University Press.

Stumpf, Rebecca M. and Boesch, Christophe. 2006. "The efficacy of female choice in chimpanzees of the Taï Forest, Côte d'Ivoire." *Behavioural Ecology and Sociobiology* 60: 749–765.

Stumpf, Rebecca M. and Boesch, Christophe. 2010. "Male aggression and sexual coercion in wild West African chimpanzees, *Pan troglodytes versus*." *Animal Behaviour* 79: 333–342.

Suzuki, Akira. 1971. "Carnivority and cannibalism observed amongst forest-living chimpanzees." *Journal of the Anthropological Society of Nippon* 79: 30.

Tomasello, Michael and Call, Josep. 2007. "Introduction: Intentional communication in non-human primates." In *The Gestural Communication of Apes and Monkeys,* Call, Josep and Tomasello, Michael (eds), 1–15. New Jersey, USA: Lawrence Erlbaum Associates.

Townsend, Simon W., Slocombe, Katie E., Emery Thompson, Melissa and Zuberbühler, Klaus. 2007. "Female-led infanticide in wild chimpanzees." *Current Biology* 17 (10): 355–356.

Tutin, Caroline E. G. 1979. "Mating patterns and reproductive strategies in a community of wild chimpanzees (*Pan troglodytes schweinfurthii*)." *Behavioural Ecology and Sociobiology* 6 (1): 29–38.

Tutin, Caroline E. G. and McGrew, William C. 1973. "Chimpanzee copulatory behaviour." *Folia Primatologica* 19: 237–256.

Van Lawick-Goodall, Jane. 1972. "A preliminary report on expressive movements and communication in the Gombe stream chimpanzees." In *Primate Patterns,* Phyllis Dolhinow (ed.), 25–85. New York, USA: Holt, Rinehart & Winston Inc.

Wiley, R. Haven. 1991. "Associations of song properties with habitats for territorial oscine birds of Eastern North America." *The American Naturalist* 138 (4): 973–993.

Zuberbühler, Klaus and Reynolds, Vernon. 2005. *The Sonso chimpanzee database.* Oxford: Oxford University Press.

A call for conformity

Gesture studies in human and non-human primates

Nicole M. Scott & Simone Pika
Center for Cognitive Sciences, University of Minnesota, Minneapolis, MN 55455, USA / Max Planck Institute for Ornithology, Humboldt Research Group "Comparative Gestural Signalling", Eberhard-Gwinner-Strasse, 82319 Seewiesen, Germany

Many advances in the study of gesture have been accomplished; however idiosyncratic methodologies and definitions have been applied for different subject pools in human and non-human animals. Since gesture use in our phylogeny dates back at least to other primates, a useful framework for furthering this study should accommodate all gesture forms regardless of species membership. Such a framework would allow for in depth comparisons also if gesture is a product of convergent evolution. The current chapter (i) provides an overview of the state of the art of gesture studies in humans and other primates, (ii) adjusts McNeill's Gesture Continuum to accommodate all gesture types by implementing Tinbergen's four "why's" and (iii) creates a comprehensive definition of gesture that satisfies studies of any animal species.

1. Introduction

Current developments in gesture studies have vastly improved our understanding of this form of communication. However, much is still being debated, especially with regard to the theory of gestural origin of language (e.g. Corballis 2002; Arbib, Liebal & Pika 2008; Armstrong & Wilcox 2007). One of the most informative techniques for unraveling the evolutionary history of the gestural modality is to study the gestural complexity of our closest living relatives, the non-human primates (hereafter primates; Tomasello 2008; Pika 2009). In order to accomplish comparative research, and therefore this task, a standard definition has to be in place. As it stands, researchers of gesture have created their own definitions, which are often tailored to their specific objectives, and therefore comparisons within, let alone between species, are only rarely accomplished (but see Call and Tomasello 2007).

The following is a discussion of recent approaches to gestural communication in humans and primates. Subsequently, to further the study of gesture in a way that accommodates all species using this modality, we construct a dimensional gesture framework by linking Tinbergen's four "why's" (Tinbergen 1963) with McNeill's Gesture Continuum (McNeill 1992). Before we offer a new direction for future studies, we introduce a dimensional model for describing gesture and provide explicit directions for its applicability. Finally, we propose a comprehensive definition of gesture, which achieves the following four goals:

1. omit any actions where the signaler achieves the goal without necessitating a reaction from the recipient;
2. remove necessity for overt intention;
3. exclude self-directed actions, which do not involve a recipient and;
4. require that the action must be recognized as meaningful for all participants.

1.1 A history of gesture studies

Perhaps William James (1890:404) would have said about gesture what he said about attention: "Everybody knows what it is". But more accurately, we can say that we mostly do not agree on what it is. Actions not typically included under the heading *gesture* include: (i) spatial and orientational movements and spacing; (ii) autonomic or nervous movements (e.g. reflexes); and (iii) affective movements (e.g. laughing and facial expressions) (for a discussion see Kendon 2004). Further, in order for an action to be considered communicative, it is required that there is an understanding between the gesturer and the receiver that said action has a meaning beyond itself; much in the same way a convention is understood that everyone else also understands it (Lewis 1968; McNeill 1992). Furthermore, the action cannot be interpreted as *gesture* unless it is presented as such that the recipient may in turn act upon it (Clark 1978). Even with a general consensus of what a gesture is, the name used to label such actions (i.e. *gesture*) varies with the group being studied. For instance, researchers of human gesture may use the term *illustrators* (e.g. Ekman & Friesen 1969), *gesticulation* (e.g. Kendon 1980) or *gesture* (e.g. Goldin-Meadow 2003; McNeill 1992). Contrarily, in primate studies only the term *gesture* is used (e.g. Call & Tomasello 2007). Concerning differences in goals, four distinct forums of gesture research appear with each creating and using its own specially catered definition. These forums emerge as studies of (1) human adults, (2) human infants, (3) apes and (4) monkeys.

1.1.1 *Human adults*

Studies on gestures of adult humans have included how gesture may function in contributing to the production of the accompanying speech (Kendon 2004) or how gesture may function to describe what may not be said with words (e.g. Goldin-Meadow 2003). Others have focused on what role gesture plays in the processes of word selection (Hadar & Butterworth 1997; Krauss, Chen & Chawla 1996) as in lexical retrieval, or formulating what is to be said (Freedman 1972). Various theories have been offered to surmise the psychological implications behind gesture production and comprehension, namely the "thinking for speaking" theory that language and thought interact to form an emergent way of thinking (Slobin 1996).

Particular emphasis has been placed on how gesture can acquire language properties and become language, as in studies of the signed languages (Armstrong, Stokoe & Wilcox 1995; Peng 1978) and of deaf individuals who have not been exposed to an accessible language model, so-called "home-signers" (for an overview see, Morford 1996). These studies have provided important findings of the resilience of language (e.g. Goldin-Meadow 1997) specifically, how gestures can be learned and comprehended as symbolic. Particularly useful in this latter endeavor, as well in the development of a gesture language system, have been studies of newly formed deaf communities (e.g. Nicaraguan sign language, Senghas, Kita & Ozyürek 2004). Importantly, for adult studies, the main gesture focus has been on movements of the hands and arms in space.

1.1.2 *Human infants*

Gesture studies of human infants have focused in detail on gestural skills of pre-linguistic human infants. These studies have gauged infants' understanding of intentions and others as communicative agents (e.g. Bruner 1981; Bates et al. 1975, 1979; Bretherton, McNew & Beeghly-Smith 1981). In addition, they investigated in detail how specific types of gesture (e.g. comment, request and emotive gestures) develop and aid in interactions with the world (e.g. Blake, McConnell, Horton & Benson 1992) and how they are changed and eventually eclipsed by speech (e.g. Blake, Vitali, Osborne & Olshansky 2005).

One important aspect of infant gesture research is the development of methods for parsing gesture from other forms of action or social behavior as well as, for example, determining the difference between similar gesture types (e.g. REACHING and POINTING; Blake, McConnell, Horton & Benson 1992). One such method is to consider only intentionally produced action (e.g. Clark 1978). Reliable parameters to infer the presence of intention are: (1) persistence to the goal; (2) means-ends dissociation (e.g. a single gesture is used in several contexts); (3) waiting of the signaler after a signal has been produced, expecting a response; (4) gaze alternation between goal and recipient; and/or (5) audience effects (sensitivity to the presence and attentional state of a given recipient) (Bates et al. 1979; Bruner 1981; Pika 2008b; Pika & Liebal this volume). Generally, the term *gesture* is used in the majority of studies to refer to movements that are performed in space and the emphasis is thus on so-called 'empty-handed' gestures (Petitto 1988; Pika 2008a).

1.1.3 *Apes*

In the earliest studies, researchers tried to investigate the communicative skills of primates by either attempting to teach them spoken language (Hayes 1951) or American Sign Language (ASL) (e.g. Gardner & Gardner 1969; Miles 1986; Patterson 1978). Later, ideographic systems were created to converse with chimpanzees and bonobos (Rumbaugh 1977; Savage-Rumbaugh et al. 1986). These investigations provided the first evidence that apes are able to communicate through true symbols (which are arbitrary to what they represent and used in a decontextualised manner; e.g. Peirce 1978). Recently, projects have steered from testing the abilities of primates to communicate with humans and have turned their focus to the natural flow of communication between conspecifics (Tanner & Byrne 1996; Call & Tomasello 2007); that is to say, unsolicited from human observers. In particular, emphasis has been shifted to the social cognitive abilities underlying gesture production and comprehension (Plooij 1984; Tomasello & Call 2007) and have provided important arguments supporting the theory of a gestural origin of language (Corballis 2002; Hewes 1973; Arbib, Liebal & Pika 2008). For instance, it has been suggested that referential gesturing, especially the POINTING gesture is unique to humans. However, a variety of studies have shown that captive chimpanzees are able to use POINTING to communicate with their human caretakers, indicating the referential potential of chimpanzees (Leavens & Hopkins 1999; Leavens, Hopkins & Bard 2005). Furthermore, Pika & Mitani (2006) have described a distinct gesture, the DIRECTED SCRATCH, used by wild chimpanzees of the Ngogo community, Uganda to specify distinct areas to be groomed.

Contrary to studies on humans, these studies have concentrated on defining gesture as movements of the head, the body and the limbs. In addition, they also incorporated repeated movements on objects or on the ape's own body, such as the gestures CHEST BEAT, BODY BEAT, SLAP GROUND and DRUM OBJECT (Call & Tomasello 2007). Furthermore, some methods used in human infant studies have been adopted, such as the criteria for establishing intentionality (e.g. Bard 1992; Pika, Liebal & Tomasello 2003; Scott 2007). More succinctly, this has been equated to goal-directed behavior by at least one of the involved interactants (see Leavens & Racine 2009).

1.1.4 *Monkeys*

Virtually nothing is known about New World monkey or prosimian gesture use (e.g. Moynihan 1967; Schaffner & Aureli 2005). However, some studies investigated the gestural abilities of Old World monkeys (Maestripieri 1999), with a special focus on social functions of gestural usage rather than distinct gesture types. For instance, Maestripieri (2005) examined the use of gestures in three species of macaques in relation to their social structures. Van Hooff (1967) compared the facial display of several monkey species in multiple contexts.

Overall, researchers of monkey gestures have applied broad descriptions of social behavior, including facial expressions and postures under the umbrella term *gesture* (e.g. Maestripieri 2005; Hesler & Fisher 2007; Wolfheim & Rowell 1972; Smuts 2002; Lairdre this volume). The focus on facial and bodily expression may be due to less emphasis being placed on the limbs: the quadrupedality of monkeys makes it difficult to free the hands for manual gestures (for similar arguments concerning apes, see Menzel 1973; Plooij 1978). Only some researchers have applied parameters used in studies of human infants and apes to investigate the cognitive complexity underlying monkey's gestural signaling (e.g. Hesler & Fischer 2007).

1.2 How do humans, apes and monkeys compare?

In this next section, we discuss the lack of compatibility of the current gesture definitions just described (see Table 1). Human gestures are often intertwined with spoken language in time (McNeill 1985). Primates on the other hand appear to use their vocalizations much more separately from gestures (e.g. Pollick & de Waal 2007; Tomasello et al. 1994), the emphasis being placed on behaviors where intentionality can be inferred. This aspect is ignored in studies of adult humans (e.g. McNeill 1992). One such instance is a distinct gesture type, the so-called BUTTERWORTHS (McNeill 1992), which is not considered to be intentionally produced by signalers. These gestures are used as 'placeholders', while the accompanying speech is mentally organized. However, on the other extreme, some researchers of monkey gestures also leave out intentionality (e.g. Maestripieri 2005; Laidre this volume).

Additional discrepancies include: (i) bias in identifying object included gesture in primate studies but not human infant studies (Pika 2008a; but see Bates, Camaioni and Volterra 1975 for an exception); (ii) categorizing gestures by modality in primate studies (e.g. Pika, Liebal, Call & Tomasello 2005) but social function in human studies (e.g. Ekman & Friesen 1969) or concentrating on only the visual modality in humans (e.g. Kendon 2004; McNeill 1992); and (iii) separating perlocutionary actions (behaviors in which communication occurs only because the receiver is adept at interpreting the behavior of the "signaler") from illocutionary acts (conventional purpose of act) in studies of infants (Bates et al. 1979). Contrary, in studies of primates the ontogeny of communicative intent has only rarely been investigated (for exceptions see Plooij 1978, 1984; Tomasello et al. 1994). Such distinction is helpful in determining the transition of ordinary actions into intentionally communicative movements and therefore important for understanding the ontogeny of primate gesture systems.

Table 1. Comparison of current gesture definitions to our proposed definition

Forum	Current definition	Lacks	Current application	Compares to other definitions	Adds to our definition
Adult humans	Movements of the hands and arms in space accompanying or replacing speech	Other body movements; independence from language	"Thinking for speaking", signed languages	Occurrence (presence or absence of listener or speech, in assistance or alliance with speech) and representation (gesture categories)	Represents a meaning beyond itself, social regulation
Infant humans	An act of intentional communication that is produced so that another may act upon it	Social regulation	Intentionality, development of communication, social manipulation	Intentionality	Restricted use of term (intentionality, directed)
Apes	Any movement of the limbs or head and body postures that appears to be communicative	Agreement between research groups	Development of communication, language properties, repertoire of species	Intentionality and flexible use, repetitive use, temporal restriction, mechanically ineffective	Mechanically ineffective, social regulation, restricted use of term (flexibility)
Monkeys	All visual, tactile, and auditory signals that are expressed through movements of the body or face, except for vocalizations	Exclusion of posture and facial expressions	Repertoire of species	Identifies morphological constraints on production	Bodily action
Revised	Any nonvocal bodily action directed to a recipient that is mechanically ineffective and represents a meaning, beyond itself, that is in part manifested by others of the social group	N/A	All species using non-vocal signals	Includes aspects from each forum of primate literature	Addresses all types of gesture use

2. Building a framework

To develop a concise framework and thereby enabling gesture researchers to talk about AND mean the same thing when using the term *gesture*, we combined two distinct approaches from two different disciplines: McNeill's Gesture Continuum and Tinbergen's four "why's".

2.1 McNeill's Gesture Continuum

McNeill's Gesture Continuum (1992, previously known as Kendon's Continuum) (see Figure 1) was developed for gesture in respect to language and constructed under the premise that by "holding medium constant, see what properties of sign are constant and what are added by conventionally structured code" (McNeill 1992: 36). It is referred to as a continuum rather than a step-wise progression since each kind of gesture grades into the next.

→ Presence of language properties increases/social regulation increases

Gesticulation → Language-like gestures → Pantomimes → Emblems → Sign languages

← Obligatory presence of speech increases/idiosyncrasy increases

Figure 1. Schemata of Gesture Continuum (adapted from McNeill 1992)

The continuum can be used to describe gesture in terms of both its function (as described by the directions of language-like properties and social regulation) and its use (as described by the directions of obligatory speech presence and amount of idiosyncrasy). Primate gestures for instance are often accompanied by vocalizations, and research shows that some populations develop idiosyncratic gestures (Pika et al. 2003, 2005; Pika & Liebal this volume). Such social regulation is also evident in the cultural variation found amongst wild chimpanzee communities (e.g. Whiten et al. 1999). These aspects of gesture use place primate gesture not within any distinct gesture type, but rather across the entire continuum.

We thus will extend the continuum further, perhaps well beyond McNeill's intention, as a representation of the evolution of gesture as communicative tool: from simple action to an organized, regulated system. In this sense the continuum may be seen to arrange gestures from a state akin to spontaneous, basic impulse towards one of derived, sophisticated orderly symbolism. Additionally, the continuum shows gestures are used flexibly and for different purposes. Although McNeill (1992) described two explanations for gesture – namely causation and function (adaptation) – by applying Tinbergen's (1963) four "whys", two additional explanations will be added (ontogeny and phylogeny).

2.2 Tinbergen's four "why's"

Tinbergen (1963) created this framework, the four "why's", to be applied by all researchers of animal behavior. The four questions were intended to focus on the development and mechanisms of behavior with respect to the benefits such actions created. Through these, Tinbergen offered two explanations of behavior: proximate explanations, which can be likened to use and ultimate explanations, which can be likened to function. The first two proximate explanations, or local causes, are *Causation* and *Ontogeny*, while the two ultimate explanations are *Function* and *Phylogeny.*

Causation explains the physical source of the behavior through the physiological mechanism and the physical stimuli, which elicit the behavior (e.g. the sympathetic nervous system's triggering the fight-or-flight response). *Ontogeny* explains the development of behavior throughout the life of the individual as well as the benefits of such behavior to the individual (e.g. critical period of language learning in humans). *Function* explains the forecasted benefits of the behavior through the current, adaptive mechanism of behavior as it has been shaped from selection pressures (e.g. social grouping as a mechanism of protection from predators). *Phylogeny* explains how the behavior developed within a given species and how the species as a whole benefits (e.g. natural selection of traits within a population which either favor species survival or do not hinder it such as the blind spot in the vertebrate eye).

2.3 Four dimensions of gesture use

Applying Tinbergen's four "why's" to McNeill's Gesture Continuum expands the continuum into a four dimensional framework. However, it is misleading to think that these dimensions expand into infinity: Tinbergen's questions also constrain the limits of the dimensions in accord with the natural constraints on the communicating individual. These dimensions are labeled according to their constraints: (1) *Mechanical,* (2) *Perceptual,* (3) *Purposeful* and (4) *Representational.*

2.3.1 *Mechanical*

The *Mechanical* dimension provides an important assessment not because it describes the morphology of the gesture, but because it indicates the morphological constraints of the signaler. For instance, a WAVE (or POINT) will not have the same morphology for a human and a chimpanzee (for an empirical assessment of differences in POINTING see Povinelli & Davis 1994). Also, there are ecological and cultural constraints to be considered both within and between species. For example, some cultures POINT with their lips (Sherzer 1972) while others would not regard this behavior as communicative without prior experience. Also, there is an increased reliance on tactile gestures in arboreal species such as orangutans and gibbons as a consequence of their ecological settings (Chivers 1976; Liebal, Pika & Tomasello 2006). Future directions in research would benefit from further investigations into mechanical restrictions. This may reveal the evolutionary mechanism of communication modality shift from visual to vocal form – especially with respect to the vast flexibility in human vocal manipulations (e.g. culture-specific phonemes) in comparison to the relatively limited repertoire of culture-specific gestures (e.g. everybody POINTS).

2.3.2 *Perceptual*

As with all other aspects of our world, we are capable of perceiving only a small fraction of the stimuli surrounding us (e.g. visual, auditory, olfactory and sensory stimuli). With this limitation, we are restricted to what elements of a signal we are capable of perceiving and/or reproducing. The *Perceptual* dimension describes how the gesture system is learned and developed with respect to the constraints of a given recipient. This helps us to understand how a particular gesture system has been acquired. Many researchers have attempted to describe the development of gesture with regards to underlying learning mechanisms use (humans: Bates et al. 1979, 1975; primates: King 2004; Plooij 1984; Tomasello et al. 1994; however see Genty, Breuer, Hobaiter & Byrne 2009). To date, no single method of development has been demonstrated as the principal method of skill acquisition for all gesture types within any single species. It thus appears there are different methods for learning different types of gestures (Pika 2008b) and it may be that humans and other primates combine and/or utilize different methods of acquisition. Although humans and primates acquire some gesture types similarly (e.g. through a process of "ontogenetic ritualization", in which a communicatory signal is created by two individuals shaping each other's behavior in repeated instances of an interaction, see Call & Tomasello 1997), it remains unknown how humans acquire some of their gesture types (e.g. tapping the air to emphasize a statement: is it the intrinsic quality of the gesture or its conventional use that propagates it?). One of the main questions therefore concerns "Does ontogeny reflect phylogeny" (a modification of the question of "does ontogeny recapitulate phylogeny" of which the general consensus is "no" – see Gould 1977 – albeit with exceptions, see Gottlieb 2002). The term "reflect" is especially used to emphasize that no re-enactment of learning exists (as if learning processes are built one upon the other in a biologically determined order, but rather that altogether new, emergent processes are developed and utilized). If the answer to this question is "no", then the development of gesture use (ontogeny) by individuals of a given species is not an informative comparison within primate phylogeny as all primates would acquire gestural use via approximately similar processes. However, if the answer to this question is "yes", then differentiation in the learning techniques could provide important information about the social and cognitive capacities of different primate species (or families, etc.). We thus possess a very useful tool to illuminate evolutionary changes within the primate lineage. In short, this dimension emphasizes the relationship between cognition and language as these likely co-evolved; thus, by understanding the learning mechanisms involved in developing a species' particular gesture system we will also gain a better understanding on the convergence (or divergence) of these abilities throughout the primate lineage.

2.3.3 *Purposeful*

The *Purposeful* dimension describes the constraint of gesture meaning with respect to the social context in which it is used. The very name of this dimension hints at its description of the purpose of a particular gesture. However, this should not be confused with the gesture's function. The purpose describes the immediate goal of the signaler, while the function describes the ultimate goal of the signaler as interpreted by the receiver[1]. It is the coordinated effort of the communication partners, which separates purpose from function. For instance, if the purpose of a given gesture is to threaten, it is classified as a "threat" gesture (e.g. a SLAP GROUND performed by chimpanzees, Scott 2007) and this is especially the case in the primate literature. So, if this "threat" gesture is used in an aggressive and threatening way, then the function is to keep the receiver at a distance. In contrast the same gesture can be used to initiate a play bout (a SLAP GROUND by the same chimpanzee, Scott 2007) and thereby attracts the receiver. In these examples, the same gesture is used for the same purpose (i.e. threat), but for different functions (i.e. repel or attract the recipient), albeit the accompanying facial expressions or other communicatory means may aid in clarifying the message. As illustrated by this example, the purpose remains relatively unchanged but through ontogeny as the individual develops physically and socially the function may change as its relationships with communication partners change.

Categorizing gestures by their social context is a common technique in primate studies (e.g. van Hooff 1973; Tomasello et al. 1994). However, human gestures are typically categorized by their function or purpose (e.g. referential

1. In an effort to avoid sounding too information theoretic: from a dynamic-systems view, perhaps this would be equal to saying the purpose is in the role of the signaler and the function is in the role of the recipient.

value, McNeill 1992). Since the social context can be defined either by purpose or function and these are not always equal, it is difficult to compare human and primate gesture use based on these categorization systems. It is thus more useful to consider the outcome of the gesture and the triangular relationship of signaler, recipient and resulting action to infer the meaning. Additionally, just as there are gestures of similar morphology but different meanings between cultures there are gestures that are morphologically shared amongst species (e.g. POINTING). With this caveat, careful consideration must be given when categorizing the purpose of gesture as there is large potential for observer bias through egocentric interpretations: with shared definitions we can avoid such biases.

2.3.4 *Representational*

Bias exists not only in *how* we consider gestures, but also in *what* we consider gestures. It is necessary to distinguish between mere reactions and intentional communication. This last dimension is useful in describing the shift from intentional action to intentional communication (e.g. Bard 1992). For instance, throwing the hands up to the face is an individual's reaction to protect her from harm, which also may be intentionally produced as a gesture of being startled. When deciding whether an action was intentionally communicative, similar approaches and reasoning should be used for all subjects of gestures as those used, historically, for human infants and for studying the origins of language (e.g. Clark 1978).

The *Representational* dimension describes the mental representation elicited by the gesture, including its subsequent interpretation. Representation is a topic of much debate (Hubbard 2007), and it is unknown to what extent animals other than humans possess this cognitive ability; namely, whether their representations are genetically fixed and attached to emotional arousal or more conscious and attached to associative memory. One view is that an experience has a biologically-fixed representation according to the function of the system that does the representing (Dretske 1995). This representation depends upon the individual's relationship with its environment and society (similar to the theory of "Umwelt", Von Uexküll 1934) and is thus constrained by both. For example, a fly may be represented in a human as a nuisance, pet, food, etc. but in a frog, it will always be represented as food – otherwise the frog may go quite hungry![2] Therefore, the

2. Even if the fly did represent food to a human (as it may to a frog) the algorithms that lead to such representation differ in accordance to the individual's specific relationship with flies, as it would with any object in their environment (Wael El-Deredy personal communication, May 2009).

representation is genetically fixed in the frog, but consciously contrived in the human (even if this choosing eventually becomes automated without conscious awareness) as our relationship with the world around us is shaped by our environment and culture. It is not yet understood how other species represent their environments.

2.4 How to apply the dimensions?

The introduced dimensions guide the decision process of gesture description (*Morphological* dimension), illustrate their social use (*Purposeful* dimension), and describe the development of the gesture system (*Perceptual* dimension). The *Representational* dimension is most useful for explaining the underlying mechanisms of phylogenic variation as the gesture systems, and their accompanying mental models, have been adapted to each species' environment. The dimensions thereby create a more complete depiction of a gesture and the different varieties that comprise this term – from speech-accompanied to those taking full communicative burden – and allow all varieties to fit within one, single definition.

To devise a definition, we combine McNeill's continuum parsing the different types of gestures with respect to function with Tinbergen's four "Whys" to parse behavior into its different mechanisms. McNeill's two-dimensional, linear model falls short of including all types of gestures. For instance, deictic gestures, as well as beats, are categorized as gesticulation but do not actually co-occur with speech but rather take the place of it for a short moment (McNeill 1992). This places them on the opposite end of the continuum. This argument is similar for the gestures of primates and pre-linguistic human infants. Our four-dimensional, comprehensive and constraining framework provides a conceptual space for arranging and comparing the different types of *gesture* within and between species. Once a deeper level of understanding is attained with respect to specific gesture types then a 4-D model can be created to visualize these comparisons. This kind of development is necessary for the advancement of comparative gesture studies.

3. Proposed definition

Using our framework, we devised a universal definition embracing all fields of gesture research that specifically addresses the four points we listed earlier (see paragraph 2). Each of these four points is implicitly dealt with by one of the dimensions in respective order. Our proposed definition is as follows: **any non-vocal bodily action directed to a recipient that is mechanically ineffective and**

represents a meaning, beyond itself, that is in part manifested by others of the social group.[3] Removing the necessity for the communicative act to be overtly intentional avoids the circular argument that results when the cognitive processes underlying gesture production are included in the definition of gesture (Müller 2005): this allows for non-trivial inclusion of the more obscure gesture types such as BEATS, BUTTERWORTHS, and those gestures produced without a visible audience.[4] Our definition, if we included vocal actions, could also encompass spoken words. It is perhaps this realization that leads some researchers to label words as gestures (e.g. Liberman, Cooper, Shankweiler & Studdert-Kennedy 1967) and yet others to postulate that spoken language developed out of a gestural communication system (e.g. Corballis 2002). A researcher who disregards intentionality of signaling may well consider any communicative effort as gesture (for an exception, see King 2004: 56–60). So what questions cannot currently be asked or adequately answered due to the current patchwork of definitions? Without a common denominator no comparisons can be made. Essentially, no questions currently can be answered confidently with respect to different populations: it is equivalent to comparing apples to oranges. With a common denominator (i.e. definition) constraint can be implemented leading to predictive power.

References

Arbib, Michael A., Liebal, Katja and Pika, Simone. 2008. "Primate vocalization, gesture and the evolution of language." *Current Anthropology* 49: 1053–1076.

Armstrong, David F., Stokoe, William C. and Wilcox, Sherman E. 1995. *Gesture and the nature of language*. Cambridge, England: Cambridge University Press.

Armstrong, David F. and Wilcox, Sherman E. 2007. *The gestural origin of language*. New York: Oxford University Press.

3. A conversation with Sherman Wilcox elicited the question of whether *spoken* language can be considered gestural as is *signed* language. If so, then "nonvocal" should be replaced by "nonverbal" or taken out completely. We would like to argue that spoken language is separate from gesture because it makes use of the vocal chords and other body parts which are not available as *visible* bodily action and thus rely on a separate modality, the auditory channel, exclusively. The next argument that can be made is that the movement and shape of the mouth could be considered gestural, but we disagree with this as the meaning of such movements is dependent entirely upon the nonvisible actions of the other body parts and can only be heard and not seen (for instance, the difference between /*b*/ and /*p*/ sounds).

4. Invisible audiences for humans include a recipient knowingly not in visual contact with the gesturer (e.g. when speaking on the phone) (Goldin-Meadow 2003). This type of audience is also evident in imaginary play, such as that of humans and primates.

Bard, Kim A. 1992. "Intentional behaviour and intentional communication in young free-ranging orangutans." *Child Development* 63: 1186–1197.
Bates, Elizabeth, Benigni, Laura, Bretherton, Inge, Camaioni, Luigia and Volterra, Virginia. 1979. *The emergence of symbols: Cognition and communication in infancy*. New York: Academic Press.
Bates, Elizabeth, Camaioni, Luigia and Volterra, Virginia. 1975. "The acquisition of performatives prior to speech." *Merrill-Palmer Quarterly* 21: 205–226.
Blake, Joanna, McConnell, Sheilah, Horton, Gayle and Benson, Nancy. 1992. "The gestural repertoire and its evolution over the second year." *Early development and parenting* 1: 127–136.
Blake, Joanna,Vitali, Grace, Osborne, Patricia and Olshanky, Ester. 2005. "A cross-cultural comparison of communicative gestures in human infants during the transition to language." *Gesture* 5: 201–217.
Bretherton, Inge, McNew, Sandra and Beeghly-Smith, Madorie. 1981. "Early persons' knowledge as expressed in gestural and verbal communication: When do infants require a 'theory of mind'." In *Infant social cognition: Empirical and theoretical considerations*, Michael E. Lamb and Lonnie R. Sherrod (eds), 333–373. Hillsdale, NY: Erlbaum.
Bruner, Jerome S. 1981. "Intention in the structure of action and interaction." In *Advances in Infancy Research*, L. Lipsitt (ed.), 41–56. New Jersey: Ablex, Norwood.
Call, Josep and Tomasello, Michael. 2007. *The gestural communication of monkeys and apes*. Mahwah, NY: Lawrence Erlbaum Associates.
Chivers, David J. 1976. "Communication within and between family groups of siamang (*Symphalangus syndactylus*)." *Behavior* 57: 116–135.
Clark, Roger A. 1978. "The transition from action to gesture." In *Action, Gesture, and Symbol*, Andrew Lock (ed.), 231–260. New York: Academic Press.
Corballis, Michael C. 2002. *From Hand to Mouth, the Origins of Language*. Princeton, NJ: Princeton University Press.
Dretske, Fred I. 1995. *Naturalizing the Mind*. Cambridge, MA: The MIT Press.
Ekman, Paul and Friesen, Wallace V. 1969. "The repertoire of non-verbal behavior: categories, origins, usage and coding." *Semiotica* 1: 49–98.
Freedman, Norbert. 1972. "The analysis of movement behavior during the clinical interview." In *Studies in dyadic communication*, Aron W. Siegman and Benjamin Pope (eds), 153–175. Elmsford, NY: Pergamon Press.
Gardner, R. Alan and Gardner, Beatrice. 1969. "Teaching sign language to a chimpanzee." *Science* 165: 664–672.
Genty E, Breuer T, Hobaiter Catherine and Byrne Richard W. 2009. "Gestural communication of the gorilla (*Gorilla gorilla*): Repertoire, intentionality and possible origins." *Animal Cognition*.
Goldin-Meadow, Susan. 1997. "The resilience of language in humans." In *Social Influences on Vocal Development*, Charles T. Snowdon and Marine Hausberger (eds), 293–311. Cambridge, UK: Cambridge University Press.
Goldin-Meadow, Susan. 2003. *Hearing gesture: How our hands help us think*. Cambridge, MA: The Belknap Press of Harvard University.
Gottlieb, Gilbert. 2002. *Individual development and evolution*. Mahwah, NJ: Lawrence Erlbaum.
Gould, Stephen. 1977. *Ever since Darwin*. New York: Norton.
Hadar, Uri and Butterworth, Brian. 1997. "Iconic gestures, imagery and word retrieval in speech." *Semiotica* 115: 147–172.

Hayes, Cathy. 1951. *The ape in our house*. New York: Harper.

Hesler, Nana and Fischer, Julia. 2007. "Gestural communication in barbary macaques (*Macaca sylvanus*): An overview." In *The gestural communication of apes and monkeys*, Josep Call and Michael Tomasello (eds), 159–196. Mahwah, NJ: Lawrence Erlbaum Associates.

Hewes, Gordon W. 1973. "Primate communication and the gestural origin of language." *Current Anthropology* 12: 5–24.

Hubbard, Timothy L. 2007. "What is mental representation? And how does it relate to consciousness?" *Journal of Consciousness Studies* 14:37–61.

James, William. 1890. *The principles of psychology*. New York: Holt.

Kendon, Adam. 1980. "Gesticulation and speech: Two aspects of the process of utterance." In *Relationships of the verbal and nonverbal communication*, Mary R. Key (ed.), 207–228. The Hague, Netherlands: Mouton.

Kendon, Adam. 2004. *Gesture: Visible action as utterance*. Cambridge, England: Cambridge University Press.

King, Barbara J. 2004. *The dynamic dance: Nonvocal communication in African great apes*. Cambridge, MA: Harvard University Press.

Krauss, Robert M., Chen, Yihsiu and Chawla Purni. 1996. "Nonverbal behavior and nonverbal communication: What do conversational hand gestures tell us?" In *Advances in Experimental Social Psychology*, Mark P. Zanna (ed.), 389–450. New York: Academic Press.

Laidre, Mark (2012). "A round-the-world study of the largest of all monkeys". In *Developments in Primate Gesture Research*, Simone Pika and Katja Liebal (eds), 111–126. Amsterdam/ Philadelphia: John Benjamins Publishing Company.

Leavens, David A., Hopkins, William D. and Bard, Kim A. 2005. "Understanding the point of chimpanzee pointing." *Current Directions in Psychological Science* 14: 185–189.

Leavens, David A. and Hopkins, William D. 1999. "The whole hand point: The structure and function of pointing from a comparative perspective." *Journal of Comparative Psychology* 113: 417–425.

Leavens, David A. and Racine, Timothy P. 2009. "Joint attention in apes and humans." *Journal of Consciousness Studies* 16: 240–267.

Lewis, David K. 1968. *Convention*. Cambridge, MA: Harvard University Press.

Liberman, Alvin. M., Cooper, Franklin S., Shankweiler, Donald P. and Studdert-Kennedy, Michael 1967. "Perception of the speech code." *Psychological Review* 74: 431–461.

Liebal, Katja, Pika, Simone and Tomasello, Michael. 2006. "Gestural communication of orangutans (*Pongo pygmaeus*)." *Gesture* 6: 1–38.

Maestripieri, Dario. 1999. "Primate social organization, gestural repertoire size, and communication dynamics." In *The origins of language: What nonhuman primates can tell*, Barbara J. King (ed.), 55–77. Santa Fe, NM: School of American Research Press.

Maestripieri, Dario. 2005. "Gestural communication in three species of macaques (*Macaca mulatta, M. nemestrina, M. arctoides*)." Gesture 5: 57–73.

McNeill, David. 1992. *Hand and Mind*. Chicago: The University of Chicago Press.

McNeill, David. 1985. "So you think gesture are nonverbal?" *Psychological Review* 92: 350–371.

Menzel, Emil W. 1973. "Chimpanzee spatial memory organization." *Science* 182: 943–945.

Miles, H. Lyn. 1986. "Cognitive development in a signing orangutan." *Primate Report* 14: 179–180.

Morford, Jill P. 1996. "Insights to language from the study of gesture: A review of research on the gestural communication of non-signing deaf people." *Language and Communication* 16: 165–178.

Moynihan, M. 1967. "Comparative aspects of communication in new world primates." In *Primate ethology*, Desmond Morris (ed.), 236–266. London: Weidenfield.
Müller, Cornelia. 2005. "Gestures in human and nonhuman primates." *Gesture* 5: 259–283.
Patterson, Francine. 1978. "Conversations with a gorilla." *National Geographic* 134: 438–465.
Peirce, Charles S. 1978. "Collected papers. Vol. II: Element of logic." In *Collected papers of Charles Sanders Peirce, Volumes I and II, Principles of Philosophy and Elements of Logic*, Charles Hartshorne and Paul Weiss (eds), Cambridge, MA: The Belknap Press of Harvard University.
Peng, Fred C. C. 1978. *Sign language and language acquisition in man and ape*. Boulder, CO: Westview.
Petitto, Laura A. 1988. "'Language' in the prelinguistic child". In *The development of language and language researchers: Essays in honor of Roger Brown*, Frank S. Kessel (eds), 187–221. Hillsdale, NY: Erlbaum.
Pika, Simone. 2009. "Our grooming cousins: Providing the link to declarative signalling?" *Studies in Communication Sciences* 9: 73–102.
Pika, Simone. 2008a. "Gestures of apes and pre-linguistic human children: Similar or different?" *First Language* 28: 116–140.
Pika, Simone. 2008b. "What is the nature of the gestural communication of great apes?" In *The shared mind*, Jordan Zlatev, Timothy P. Racine, Chris Sinha and Esa Itkonen (eds), 165–186. Amsterdam/Philadelphia: John Benjamins Publishing Company.
Pika, Simone and Liebal, Katja. 2012. "Introduction. Developments in primate gesture research." In *Developments in Primate Gesture Research*, Simone Pika and Katja Liebal (eds), 1–11. Amsterdam/Philadelphia: John Benjamins Publishing Company.
Pika, Simone, Liebal, Katja, Call, Josep and Tomasello, Michael. 2005. "The gestural communication of apes." *Gesture* 5: 41–56.
Pika, Simone, Liebal, Katja and Tomasello, Michael. 2003. "Gestural communication in young gorillas (*Gorilla gorilla*): Gestural repertoire, learning and use." *American Journal of Primatology* 60: 95–111.
Pika, Simone and Mitani, John. 2006. "Referential gestural communication in wild chimpanzees (*Pan troglodytes*)." *Current Biology* 16: 191–192.
Plooij, Frans X. 1978. "Some basic traits of language in wild chimpanzees?" In *Action, Gesture and Symbol*, Andrew Lock (ed.), 111–131. London: Academic Press.
Plooij, Frans X. 1984. *The behavioural development of free-living chimpanzee babies and infants*. Norwood, NJ: Ablex.
Pollick, Amy S. and de Waal, Frans B. M. 2007. "Ape gestures and language evolution." *Proceedings of the National Academy of Sciences of the United States of America*. 104: 8184–8189.
Povinelli, Daniel J. and Davis, D. Richard. 1994. "Differences between chimpanzees (*Pan troglodytes*) and humans (*Homo sapiens*) in the resting state of the index finger: Implications for pointing." *Journal of Comparative Psychology* 108: 134–139.
Rumbaugh, Duane M. 1977. *Language learning by a chimpanzee. The Lana project*. New York: Academic Press.
Savage-Rumbaugh, E. Sue, McDonald, Kelly, Sevcic, Rose A. Hopkins, William D. and Rupert, Elizabeth. 1986. "Spontaneous symbol acquisition and communicative use by pygmy chimpanzees (*Pan paniscus*)." *Journal of Experimental Psychology* 115: 211–235.
Schaffner, Colleen M. and Aureli, Filippo. 2005. "Embraces and grooming in captive spider monkeys." *International Journal of Primatology* 26: 1093–1106.
Scott, Nicole M. 2007. *Communicating Dominance: Chimpanzee use of gestures*. M-Res Dissertation, Roehampton University, Roehampton, UK.

Senghas, Ann, Kita, Sotaro and Özyürek, Asli. 2004. "Children creating core properties of language: Evidence from an emerging sign language in Nicaragua." *Science* 305: 1779–1782.
Sherzer, Joel. 1972. "Verbal and nonverbal deixis: The pointed lip gesture among the San Blas Cuna." *Language and Society* 2: 117–131.
Slobin, Dan. 1996. "From "thought and language" to "thinking for speaking"." In *Rethinking linguistic relativity*, John J. Gumperz and Stephen C. Levinson (eds), 70–96. Cambridge, England: Cambridge University Press.
Smuts, Barbara. 2002. "Gestural communication in olive baboons and domestic dogs." In *The cognitive animal: Empirical and theoretical perspectives on animal cognition,* Marc Bekoff, Colin Allen and Gordon Burghardt (eds), 301–306. Cambridge, MA: MIT Press.
Tanner, Joanne E. and Byrne, Richard W. 1996. "Representation of action through iconic gesture in a captive lowland gorilla." *Current Anthropology* 37: 162–173.
Tinbergen, Niko. 1963. "On aims and methods of ethology." *Zeitschrift für Tierpsychologie* 20: 410–433.
Tomasello, Michael. 2008. *Origins of Human Communication*. Cambridge, MA: MIT Press.
Tomasello, Michael and Call, Josep. 1997. *Primate Cognition*. New York: Oxford University Press.
Tomasello, Michael, Call, Josep, Nagell, Katherine, Olguin, Raquel and Carpenter, Malinda. 1994. "The learning and use of gestural signals by young chimpanzees: A trans-generational study." *Primates* 35: 137–154.
van Hooff, Jan A. R. A. M. 1967. "The facial displays of the Catarrhine monkeys and apes." In *Primate ethology*, Desmond Morris (ed.), 7–68. London: Weidenfield.
van Hooff, Jan A. R. A. M. 1973. "Structural analysis of behaviour in chimpanzees." In *Social Communication and Movements: Studies in interaction and expression in man and chimpanzee,* Mario von Cranach and an Vine (eds), 75–162. London: Academic Press.
Von Uexküll, Jakob. 1934/1957. "A stroll through the worlds of animals and men: A picture book of invisible worlds." In *Instinctive behavior: The development of a modern concept*, Claire H. Schiller (ed.), 5–80. New York: International Universities Press.
Whiten, Andrew, Goodall, Jane, McGrew, William C., Nishida, Toshisada, Reynolds, Vernon, Sugiyama, Y., Tutin, Caroline E. G., Wrangham, Richard W. and Boesch, Christophe. 1999. "Cultures in chimpanzees." *Nature* 399: 682–685.
Wolfheim, Jaclyn H. and Rowell, Thelma E. 1972. "Communication among captive talapoin monkeys (*Miopithecus talapoin*)." *Folia Primatologica* 18: 224–255.

Cognitivism, adaptationism and pointing

Timothy P. Racine
Simon Fraser University

I analyze the conceptual foundations of Tomasello and colleagues' influential theory of the origins of POINTING gestures in an attempt to clarify its aims and limit its scope. After some preliminary remarks, I consider the postulated cognitive mechanisms that are claimed to be the developmental basis for the sharing of intentional states with others, and then turn to the postulated evolutionary mechanisms that are said to serve as a basis for cognitive abilities. I discuss the strengths and limitations of this account in terms of its parent meta-theoretical positions of cognitivism and adaptationism respectively.

As the chapters in this edited volume and a variety of contemporary studies should suggest, scientists have learned quite a bit more recently about the gestural and related social cognitive capacities of non-human primates, especially great apes (henceforth, apes) (for recent reviews, see Leavens & Racine 2009; Pika 2008). However, by comparison, an aspect of this research that has been neglected is the analysis of the frameworks within which such capacities are conceptualized and investigated (Andrews 2005; Boesch 2007; Glock 2009; Johnson 2001; Leavens, Hopkins & Bard 2008; Racine in press). Therefore, rather than provide an additional review of primate POINTING gestures, my intention is to provide more an archaeology of the conceptual foundations of much of this area of research as exemplified through an analysis of what is probably its most influential theory. Conceptual frameworks differ in their suitability for application to particular problems, and the main point I wish to make is that cognitivist and adaptationist approaches do not really tell us very much about the origins of POINTING. This is mostly because their redescriptions of the phenomena are not a replacement for a careful explanation of how or why it is that POINTING actually develops in humans, and how and why, for the most part, it does not in other species.

Although the application of any conceptual framework runs the risk of distorting the phenomena that its users are trying to understand, I attempt to show in this paper that cognitivist and adaptationist approaches to POINTING project particular framework-internal conceptual difficulties that are ill-suited to the object of inquiry. The framework-internal problems that ramify through explanations of POINTING can be characterized as follows. First is the cognitivist practice of an in principle separation of mind and behavior, and the related demand for representational capacities from domains that do not require, or in either case cannot empirically demonstrate, them. Second is the adaptationist practice of noting a present adaptive behavior or species difference and reducing this difference to adaptive radiation and the presumed set of genes that underwrites it, which is again neither required nor easily resolved empirically.

Although cognitivism is a familiar term, what I have in mind by adaptationism is perhaps most helpfully understood as *the belief that one has explained a given phenotype or phenotypic difference by stating that it has been caused by an adaptation without providing additional proof that this is indeed the case.*[1] (Additional proof, by definition, cannot be the redescription of the phenotype or phenotypic difference.) Similarly, the presence of "mental states" for a given behavior can be empirically supported by appropriate modes of evidence, minimally, in the case of claims of representation of an internal model of a physical or social object, that agents do actually represent the given stimulus. Unfortunately, representation is typically stipulated rather than empirically supported in the cognitivist accounts. Therefore, one could say in parallel that cognitivism is *the belief that mental states have caused a phenotype or phenotypic difference without satisfying any additional burden of proof.* (Additional proof, by definition, again cannot be the redescription of the phenotype or phenotypic difference.) The underlying problem in both belief systems tends to be that mental states and adaptations are taken in far too literal a manner; they are at best placeholders for some future explanation that has yet to be delivered.

In this chapter, I first make some introductory remarks concerning mental states and adaptations and then analyze the conceptual foundations of the latest version of Michael Tomasello and colleagues' influential theory of the origins of POINTING gestures in an attempt to clarify its aims and limit its scope. I consider the postulated cognitive mechanisms that are claimed to be the developmental basis for the sharing of intentional states with others, and then turn to the postulated

1. Given space concerns, my goal in this chapter is not to examine in detail the conceptual foundations of cognitivism, but rather to see how cognitivism (and adaptationism) is playing out in a dominant theory of human/non-human social cognition. There is a variety of other relevant work that goes into the problems of cognitivism in the sort of research in more detail and I direct interested readers to, for example, Leudar and Costall (2004), Sharrock and Coulter (2004), and ter Hark (2006). The same holds for adaptationism (e.g. Gottlieb 2002, Lewontin 2000, Lickliter 2008).

evolutionary mechanisms that are said to serve as a basis for these cognitive abilities. I discuss the strengths and limitations of this account in terms of its parent meta-theoretical positions of cognitivism and adaptationism.

1. Of mental states and adaptations for mental states

Debates regarding the meaning of POINTING in humans and apes are often framed in terms of mutually-exclusive polarities that are difficult to resolve on scientific or logical grounds. For example, researchers often puzzle over whether young humans or apes understand others in terms of their surface behavior *or* their underlying mental states (for a recent example, see Tomasello, Carpenter & Liszkowski 2007; and the commentaries thereupon by D'Entremont & Seamens 2007; Gómez 2007; Southgate, van Maanen & Csibra 2007, see also e.g. Penn & Povinelli 2007). This is not a contrast that is inherently problematic because it is obviously possible to act on the basis of surface behavior, which I would likely do if I came across a bear on the path that I walk to work. However, within cognitivism, and relatedly as we will see adaptationism, this sort of reasoning is stipulated in a general sense and introduces the problem of gluing mind and behavior back together. These sorts of inner-outer/mind-behavior polarities are therefore not helpful in a stipulative sense. They have also contributed greatly to the polarization that exists in the explanation of human and ape POINTING gestures.

Although contemporary scientists are rarely recalcitrant Cartesians and are therefore using such polarities as terms of art to make particular theoretical and empirical claims, I find myself in the position of having to translate their claims into a language that I can understand; a language that preserves the intrinsic relation between mind and behavior. In my view, such a language is largely absent in the modern debates about the gestural and social cognitive capacities of young humans and apes. Thus, although the intuition involved in theorizing about surface behaviors versus underlying mental states is capturing something, namely that it is possible to act or react with thinking, many contemporary researchers seem to have forgotten that mental state concepts are applied on the basis of behavior. Particular sorts of behavior, including linguistic behaviors, express particular mental states. To think otherwise is to conceive of mental states in misleadingly concrete and reified terms. As Wittgenstein (e.g. 2009) was keen to remind us, whereas this is fine in our everyday speech, it is a poor basis for scientific explanation.

Mental states and adaptations are best thought of as abstractions; they do not describe physical events in the way that say under Newtonian physics the mass of an object is a function of its force and acceleration.[2] I am not meaning to claim that all concepts are not abstractions because the concepts of the Physicist are as well. The point is that this does not mean that adaptations or mental states are *less accurate* descriptions of physical events; rather, they are not, strictly speaking, descriptions of physical events at all. Although the genes that might be taken (I will argue misleadingly) to cause a given adaptation, or the neural circuitry that might be taken to cause a particular mental state, might involve physical objects, the mental state or adaptation clearly does not. After all, one cannot smash shared intentions or an adaptation for shared intentionality in a particle accelerator. Physicists typically seek to explain relations between various law-governed forces and the behavior of physical objects; psychologists often seek to explain relations between intentions and intentional behavior, or between adaptations and adaptive behavior. In both cases, it is misleading to (simply) think about the inner. Although conceptual analysis, which involves paying attention to the meanings of the scientific concepts under study and examining the coherence of claims that come from theory or empirical work, is an essential part of science in general, it is neglected in Psychology and allied disciplines. This is problematic given the highly interpretative nature of psychological explanation (Machado & Silva 2007; Racine & Müller 2009).

Because their approach is explicitly cognitivist and adaptationist and therefore exemplifies the chief points I wish to discuss in this chapter, I now turn to the theory of Tomasello and colleagues. The term "shared" or "joint" intention has become an important concept at work in this theory, and it turns on the meaning of the technical concept "shared intentionality" that they have introduced in the most recent theoretical work (Tomasello 2008, 2009; Tomasello & Carpenter 2007; Tomasello, Carpenter & Liszkowski 2007; Tomasello, Carpenter, Call, Behne & Moll, 2005). This use of this term and related ones is explicitly cognitivist and adaptationist and a goal in this chapter is to suggest that this largely underlines the power explanatory of their theory. However, as my colleagues and I have documented elsewhere (e.g. Leavens & Racine 2009; Racine & Carpendale 2007; Racine, Leavens, Susswein & Wereha 2008/2010; Racine, Wereha & Leaven in press; Wereha & Racine in press), Tomasello is by no means unique in his use of these meta-theories and although my focus on his and his colleagues' theory

2. I say mostly because it makes sense to think of certain psychological concepts as states. For example, I can time a sensation or certain emotional states and they can be interrupted by attentional shifts and the like (for more detail concerning these aspects of the use of such psychological concepts, see e.g. Bennett & Hacker 2003; Racine in press; Racine & Carpendale 2007; Susswein & Racine 2008).

(Tomasello 2008, 2009; Tomasello & Carpenter 2007; Tomasello et al. 2005, 2007) is a critical one, I aim to illustrate more general concerns.

2. The origins of pointing in humans and apes

I have reviewed the logic and source of the concept of shared intentionality elsewhere in greater detail (Racine in press). In brief, my chief points in that paper were to show that (a) the concept is used in a way that is not entirely faithful to its use in the philosophy of action literature from which it was appropriated; (b) the concept is explicitly cognitivist and that Tomasello and his colleagues (Tomasello 2008, 2009; Tomasello & Carpenter 2007; Tomasello et al. 2005, 2007) cast their explanation in terms of dichotomies that are not appropriate for the explanation of joint attentional capacities; (c) their claims were nor strictly-speaking empirical ones but more redescriptions of the phenomenon in cognitivist terms. What I add on this point in the present paper is the suggestion that Tomasello and his colleagues engage in some serious overreach when they employ the concept as an explanation for POINTING. That is, although the back-and-forth sorts of cooperative activities that they juxtapose in typically developing human young children and captive apes might mesh quite well with what Searle (1990) and others upon whom they draw have described as joint intentions, joint commitments, and so on, there is no (empirical) reason to explain POINTING in the same way.

Joint attention and certain forms of POINTING gestures have existing meanings. As is well known, Bates and her colleagues (Bates, Camaioni & Volterra 1975) introduced the celebrated distinction between protoimperative and protodeclarative forms of gesture, the former being a request for an object or some state of affairs, the latter as a request to share attention around an object or state of affairs as an end in itself. Bates et al. considered these gestural forms to show a capacity for what theorists often call 'joint attention' and explained both with the same mechanism. That is, these gestures show a rudimentary grasp of the concept 'attention' (Racine & Carpendale 2007). Since Bates and colleagues seminal article, many in the field have taken declarative forms of communication to evince a more advanced form of social cognition than do imperative forms of communication (Pika 2008). However, as Tomasello and his colleagues' (2007) recent attempt to defend a so-called rich interpretation of human POINTING gestures would suggest, this has proven to be fairly controversial and hard to resolve (see commentaries published with the article). Although Tomasello (2003) claims that instrumental POINTING may arise in humans from ontogenetic ritualization *or* understanding the mental states of others, this is a bit of a shell game. When we say that POINTING means an agent understands attention or intention, we are not making a claim about what causes that understanding; we are making a claim about whether the grounds for attribution are satisfied or not.

How do we account for the finding that captive apes POINT? One way is to say that because apes and humans might POINT because of different causal factors (e.g. conditioning versus simulation respectively) apes do not *really* POINT (e.g. Tomasello 2006), but the problem here is that the meaning of the behavior is typically independent of what causes it. We did not change the meaning of the concept 'inheritance' when it was found out that genes were the "causal mechanism" involved. The same logic applies to 'joint attention' because if POINTING is said to index a rudimentary level of psychological awareness, then it does not matter how it is caused. Now, when viewed from the perspective of cognitivism of course this will seem a strange or obviously false claim because the meaning of a concept like attention or intention is claimed by cognitivists to be some internal state or the other in an organism. But although this reasoning is common, it is mistaken; when we make a claim about an agent attending or intending we are making a claim about what they do in some relevant situation. Such an attribution will presuppose certain capacities in the agent so described, but *it is not those capacities that define the meaning of the concept*. Now, this is not an entirely hard and fast rule and sometimes we might adjust the meaning of a concept because we discover something about how it is caused, what it is composed of, and so on, but if these empirical issues are not kept separate from conceptual (definitional) ones confusion is the typical outcome.

Another way to account for the finding that captive apes POINT is to distinguish forms of POINTING and claim that they index different sorts of mental operations. In this view imperative POINTS do not (necessarily) show an understanding of attention and intention, but declarative POINTS do. This would be a nice heuristic and would give us a principled basis for crediting (declaratively) POINTING typically developing humans with a clear early grasp of mind while denying such a grasp to (imperatively) POINTING apes and children with autism. This is essentially what Tomasello and colleagues defend in their 2007 article (and which Tomasello endorses in two recent books, Tomasello 2008, 2009). However, Tomasello et al. (2007) needed to resort to quite a bit of ad-hockery to make it all work. In brief, they took Bates and colleagues' (1975) original classification system and reclassified what they (Liszkowski, Carpenter, Striano & Tomasello 2006) previously termed "informative pointing" as a second form of declarative POINTING, and also introduced a new concept called an 'indirect imperative'. Essentially, this new form of imperative POINTING, of which they themselves admitted was difficult to find pre-linguistic examples, is an example of a uniquely human imperative form of POINTING. Although their article presents a "New Look at Infant Pointing", this newness that afforded them an additional way of making a clear separation between the social cognitive skills of humans and our nearest living relatives was mostly due to the taxonomic framework that they introduced. However, the general form of their claim is not new; Tomasello and colleagues have been making similarly forms of this argument for over 15 years (see e.g. Tomasello 1999).

I want to focus now in more detail on the notion of "shared intentionality" and after so doing I will explain why I think that the explanation of the social cognitive differences between apes and young humans in terms of an adaptation (or set of adaptations) for shared intentionality is also more old wine in new bottles.

3. Pointing, collaborative activities, and shared intentionality

I begin this section by making explicit four conceptual points, which I then apply to the theory of Tomasello and colleagues.

1. The meaning and cause of an activity like POINTING are independent of each other. Thus, although a researcher might have grounds to argue for differential causation, differential meaning is not possible without some new definitional stipulation.
2. Whether or not particular mental states/images accompany POINTING is independent of whether such states/images were causally required in order for that behavior to first occur. Thus, if particular mental states/images do sometimes occur while POINTING, there is no a priori reason to assume that they need to occur during each instance of it.
3. The conception of independent realms of outer behavior and inner images/states in the head is not warranted by what has been observed. Rather, it is a particular way of thinking about the relation between mind and body that was made most explicit by Descartes.
4. As a consequence of (1–3), what might appear to be empirical claims about POINTING may turn out not to be. Also, theories based upon such considerations would not be based upon scientific discoveries, although they might appear to be.

I now consider the recent theory of Tomasello and colleagues (Tomasello 2008, 2009; Tomasello & Carpenter 2007; Tomasello et al. 2005, 2007) in general terms in light of these conceptual points.[3] As I understand it, the essential elements of their argument are as follows.[4]

a. Human infants POINT because of an evolved cooperative motivation to engage in shared intentionality with others.
b. When humans begin to POINT, it is because of these causes (an adaptation or set of adaptations, and experience of their own intentional mental states and the simulation of those of conspecifics).
c. Apes do not POINT in the wild.
d. Captive apes POINT only for instrumental reasons and apes cannot declaratively POINT. (These are not the same thing, although they are often taken to be.)
e. Captive apes POINT because of conditioning.
f. There is a distinction between dyadic/ritualized forms of understanding and triadic ones that involve role-reversal/second-order representation.

Although the commitments evident in the cognitivist and adaptationist conceptions of human POINTING that are presented in (a) and (b) will be presently analyzed, (c–f) are not inherently problematic and also seem to *empirically* support the cognitivist and adaptationist claims. However, Leavens and colleagues (2008) introduced the notion of a 'Referential Problem Space' to show why apes might POINT in captivity, but not in their natural habitats. Leavens, Racine and their colleagues and others (e.g. Bates, Camaioni & Volterra 1975; Moore & Corkum 1994) have noted that (d) does not follow because all forms of POINTING may be ontogenetically instrumental even if they become functionally distinct. This issue is mostly ignored because researchers focus on the functionally distinct uses of POINTING; in

3. I say recent to distinguish this from the earlier theory that is summarized in Tomasello (1999). The essence of the early theory was that human infants understand aspects of the mental life of others through processes of simulation and analogistic extension. It was in this sense a more individualistic theory than the more recent one that emphasizes shared mental states that are themselves primitive. However, the form and conclusions of the argument are identical in both cases because they are causal psychological redescriptive accounts that are used to argue that apes and human infants differ radically in the form of their mental lives. Although there are probably many senses in which infant humans and apes differ in their "mental lives", the problem is that the evidence and theory put forward by Tomasello et al. do not speak in favour or against a range of such possibilities. In large part this is due to their meta-theoretical commitments in my view.

4. Readers interested in a sampling of quotations from the Tomasello group's work to justify this characterization should consult Racine (in press). In the present paper, which has a quite different scope and goal, I have included a concise summary of their position rather than a further list of quotes. I encourage the reader who thinks I am reading things in to consult the original sources.

Susswein & Racine's (2008) terms, they collapse over causal and definitional issues. Instead, this matter plays out in the field in general and in the commentaries published with the paper of Tomasello et al. (2007) in particular as a debate between rich and lean views of infant POINTING, with Tomasello and colleagues explicitly championing a rich view.

This seems to be the first clear Cartesian move in the game. Rich and lean interpretations of POINTING share the assumption that the issue is whether agents understand "mere behavior" versus "the underlying mental states" (see Racine, in press). However, because Tomasello and colleagues focus the debate on non-instrumental ("functionally distinct") forms of POINTING, it is forgotten that conditioning is a plausible mechanism for the emergence of any form of POINTING, human or ape (something Tomasello [2003] himself grants). Given how most researchers see the meaning of ape POINTING and what causes it, the fact that there is no evidence of role-reversal in apes, at least regarding their performance in social games as investigated by Tomasello et al. (e.g. Warneken, Chen & Tomasello 2006) closes the matter for most. However, researchers cannot find empirical evidence of shared intentionality in other apes if the ability to exchange social roles requires a human rearing environment for the development of such skills.

Let us assume though that apes (innately) can not really POINT, or do not really understand joint intentions, or only understand intentions in competitive rather than cooperative interactions. Shared intentionality, as used by Tomasello and his colleagues (Tomasello 2008, 2009; Tomasello & Carpenter 2007; Tomasello et al. 2005, 2007), is a taxonomic construct that is meant to single out collaborative activities that humans do not share with apes. For example, there is evidence that playing games involving these sorts of coordinated exchanges (for example 18- and 24-month-old human children spontaneously bouncing a wooden block on a small trampoline with an experimenter) may be unique to humans and may well turn on the respective motivations of humans (reared in typical Western human families) versus (zoo-reared) apes (Tomasello & Carpenter 2005; Warneken et al. 2006, but see Pika & Zuberbühler 2008). However, perhaps the uninitiated will find it surprising to learn that these sorts of social games have much in common with POINTING exchanges; it seems fair to say that these activities are structurally quite different. It is the theoretical apparatus that Tomasello and colleagues use that mostly unites them.

These sorts of collaborative activities do conform fairly well though to the philosophy of action literature that Tomasello and his colleagues have drawn on in past few years. This literature attempts to characterize activities that as Searle (1990) says are primitive and not reducible to individual states. For example, "The football team has the sense of 'us against them' and it has that sense against the sense of the larger us of 'teams playing the game'" (Searle 1990: 410). However, one might wonder on what basis one could attribute such a capacity to nonlinguistic agents, be they human or ape. It may be telling that although there are occasional nonlinguistic examples used in the philosophy of action literature, there are never any pre-linguistic ones. In either case, what has this to do with POINTING or even intentional behavior? Although collaboration of this sort, like most behavior that agents engage in, *is* intentional behavior, collaboration is not defined by the intentions of its actors. The reason we say that two agents are engaged in a cooperative game is that they are playing the game (cooperatively); it is presupposed that they intend to play the game. Unlike intention, which is incidental (or external) to the meaning of the activity, whether agents are cooperating is settled by how they interact. So, we have grounds for saying that such activities are cooperative or not. However, we have no grounds for saying whether they are intentional or not; it is part of the logic of the concept 'cooperative' that it be intentional. One can not cooperate by accident; that is not what we mean by cooperation. That is not how this concept works. (Although one might well say that one *meant* to cooperate, but this would be another matter.)

Does it make sense to think of cooperation/collaboration as a defining feature of POINTING? Tomasello (2008, 2009) certainly thinks so, but I believe he is wrong. If one can use a POINTING gesture cooperatively or not, then cooperation cannot be intrinsic to the concept of POINTING. However, Tomasello does *stipulate* that POINTING is inherently cooperative for humans and performed for different motivations when it occurs in apes. Now, although one can stipulate that cooperative behavior inherently involves intentions to cooperate, this is a meta-theoretical commitment. Similarly, such commitments explain why Tomasello and colleagues (Tomasello 2008, 2009; Tomasello & Carpenter 2007; Tomasello et al. 2005, 2007) use shared intentionality, successfully or otherwise, not as a taxonomic but rather a causal explanatory construct. This follows from their uncritical acceptance of the representational theory of mind that is common to both Tomasello's work and the philosophy of action literature upon which he draws.

I now turn to the proposed evolutionary roots for a quite unsound concept. My main point is to suggest that the adaptationist stance upon which Tomasello and colleagues (Tomasello 2008, 2009; Tomasello & Carpenter 2007; Tomasello et al. 2005, 2007) rely is out of step with innovations in evolutionary theory and discoveries in evolutionary biology in the past 15 years. In my view, this is a double liability in their theory of the origins of POINTING gestures and by parity their denial of such capacities in apes. However, to repeat an important point, I am *not* claiming that apes *are* humans or that they do not differ in important ways in their mentality. What is critical but often forgotten in comparative gesture research is that the grounds for attributing a variety of mental concepts *are* linguistic and therefore can be applied to neither infants nor apes. Thus, when we apply such concepts it is typically in an attenuated and rudimentary sense. But this is not to say that the exact set of concepts would be equally applicable to pre-linguistic and non-linguistic agents; this work needs to be done on a case-by-case basis. In the case of intentions and attention though there are no logical reasons to differentially apply these concepts. Thus, with respect to the rudimentary understanding of attention and intention, apes and young humans are on equal footing. Remember too that this is not a claim that they are caused by the same mechanisms; this is an empirical issue. Whether they are caused by internal psychological causal states called intentions however is not an empirical issue, although it is often taken to be. Such is the large shadow cast by the representational theory of mind over the present generation of researchers.

4. Adaptationism and adaptations for shared intentionality

Fossils leave little direct information about the psychological capacities of our forebears. One can look for genetic markers for homologous traits, or make inferences about capacities based on reconstructive anatomy of skulls, but a more typical way is to study extant species that are closely related to humans, such as the common chimpanzee (*Pan troglodytes)*. One form of explanation that has come under considerable strain in recent years proposes that cognitive adaptations underlie psychological abilities. Oftentimes, the target is so-called Evolutionary Psychology (EP; e.g. Buller 2005), but developmental systems approaches to evolution (e.g. Gottlieb 2002; Lewontin 2000; Lickliter 2008) are critical of *any* approach that relies on the Neo-Darwinian Modern Synthesis meta-theory of evolution. A thorough appraisal of EP, systems-based views of evolution, or related developments in evolutionary developmental biology (evo-devo; see e.g. Amundson 2005) would take us very far afield, but it is ironic that Tomasello (2003) has been critical of EP explanations of language and has argued strenuously that the notion of a language instinct is quite vacuous. Enter the shared intentionality instinct.

To be fair, some version of such an argument could be true, but if developmental systems views of evolution are correct, then postulating an adaptation for shared intentionality is far from an explanation of cooperative behavior, whether or not one accepts that POINTING falls under the same concept. This is because systems views of evolution conceive of a phenotype as an emergent bidirectional developmental outcome of, yes genetic, but also epigenetic effects happening at multiple levels of neural, behavioral, and environmental levels. A critical level of influence seems to be the rearing history of the organisms that are compared in experimental studies (Boesch 2007; Leavens, Hopkins & Bard 2008, Leavens Racine & Hopkins 2009; Racine, Wereha and Leavens in press; Savage-Rumbaugh, Shanker & Taylor, 1998). Further, although the science of genomic imprinting is relatively young, the extent of the role that transmission of epigenetic factors might play in phenotype is relatively unknown (Jablonka & Raz 2009). Lewontin (2000) has also demonstrated the importance of what is often dismissed as "developmental noise" for the realization of phenotype; quantum uncertainty may work all the way down when we are dealing with molecules at the level of a cell nucleus. However, the developmental systems approach is not a filling of parameters on an as yet incomplete regression equation; it is more about changing how we think about the determinants of evolutionary change. The idea that researchers can simply control for the experimental histories of test organisms is completely at odds with such a view.

Given that Tomasello and colleagues (Tomasello 2008, 2009; Tomasello & Carpenter 2007; Tomasello et al. 2005, 2007) do not specify otherwise and speak in an unqualified manner about adaptations(s) for shared intentionality, I can only assume that they intend to couch their argument within the Neo-Darwinian metatheory that systems and evo-devo researchers have increasingly begun to question. The issue at hand though is not so much with the Neo-Darwinian framework itself, but rather with its applicability to explaining development (Lewontin 2000; Lickliter 2008). The elephant on the table is that in the formation of the framework that underlies the modern synthesis of Mendelian genetics and Darwinian natural selection, reliable development was assumed so that developmental variation could be excluded. As a result, developmental processes were understood, deterministically, to be driven by genes, and evolution was seen to work chiefly through natural selection acting on heritable variation in mutation and genetic recombination and segregation. The acknowledgement of the effects of genetic drift and gene flow did not upset this classical picture because the concern was still with changes in allele frequencies in populations. However, developmental and comparative psychologists necessarily have to deal with what classical geneticists chose to ignore to make their statistical models work. If this is true, it would seem that in their theory Tomasello and colleagues (Tomasello 2008, 2009; Tomasello & Carpenter 2007; Tomasello et al. 2005, 2007), and to a large extent developmental and comparative researchers in general, must move beyond the population genetic view of development as largely directed by genetic activity.

5. Concluding remarks

Despite the growing literature on primate gestural and social cognitive capacities, what is largely absent in this body of work is an appreciation of how to understand its scope and limitations. The empirical work of Michael Tomasello and his colleagues has taught us a great deal about the communicative and social cognitive skills of the great apes, but the theoretical overlay that much of their work is presented with is often mired by ad-hockery and unacknowledged philosophical presuppositions. I have tried to show that the cognitivist and adaptationist meta-theories upon which Tomasello and colleagues (Tomasello 2008, 2009; Tomasello & Carpenter 2007; Tomasello et al. 2005, 2007) seem to quite self-consciously rely are problematic for explaining something like the origins of POINTING. On the one hand, they often trade in cognitivist metaphor without seeming to recognize that the claims only make sense within a cognitivist framework. On the other hand, they present what they claim is evidence for uniquely human adaptations to share intentions that are meant to simultaneously explain how and why human infants develop their distinctive gestural and social cognitive skills and how and why apes do not. However, these claims are really little more than a reformulation of Tomasello's (1999) position that humans have an adaptation for culture called 'joint attention'.

If the problems with Tomasello's and similar theories are directly attributable to their meta-theoretical commitments, what adaptationism and cognitivism deeply have in common is they are both based on the insertion of reified reductions of complicated processes into simple explanatory concepts such as 'mental state' or 'adaptation'. To be clear, however, there is no incoherence in creating simplified models of complicated processes that a research community can slowly build up a more detailed understanding about. The problem, a fatal one in fact, is that researchers tend to take these metaphors at face value and think they are actually assembling an understanding of something they cannot presently quite apprehend. They take themselves to be studying literal mental states and adaptations, albeit a manifestation of them as the case may be, and so on. However, although theories are often predicated on the existence of such entities, there are literally no such things. If we think there are, we have been misled by our use of concepts and sent ourselves on a wild goose chase. This, of course, should not be taken as the ridiculous statement that evolutionary or psychological processes do not really exist; of course they do. The point is there are no adaptations or mental states that we can point to in the way that we can point to genes or neurons. If we posit the existence of the former as a causal force responsible for a behavior of interest, we have misunderstood their nature.

Now, maybe Tomasello and others (e.g. Baron-Cohen 1995; Leslie 1991) are right that the experiences of our hominid ancestors stamped these skills into our cognitive repertoire, but despite the wealth of studies cited by Tomasello and his colleagues (Tomasello 2008, 2009; Tomasello & Carpenter 2007; Tomasello et al. 2005, 2007) in support of their theory, they have not really presented any independent evidence that suggests this is the case. At the end of the day, they seem to be claiming that the reasons that humans do x while apes do not is because we/they were born that way. That may well be, but this is not an explanation that is likely to pass the sniff test with a growing number of biologists. Also, if that really is the claim, then the neo-Cartesian appeals to constructs such as shared intentionality are unnecessary and probably a bit distracting. The shortcomings of this theory in my view are a result of its commitments to the representational theory of mind and population genetic traditions, both of which while still having their adherents seem to be taking on more water. The time for a sea change may be at hand.

Acknowledgements

Preparation of this chapter was supported by a Standard Research Grant from the Social Sciences and Humanities Research Council of Canada. I thank Katja Liebal, Simone Pika, Tyler Wereha, and three anonymous reviewers for comments on previous drafts.

References

Amundson, Roy. 2005. *The Changing Role of the Embryo in Evolutionary Thought: Roots of Evo-Devo*. New York: Cambridge University Press.

Andrews, Kristin. 2005. "Chimpanzee theory of mind: Looking in all the wrong places?" *Mind and Language* 20: 521–536.

Baron-Cohen, Simon. 1995. *Mindblindness: An Essay on Autism and Theory of Mind*. Cambridge, MA: Cambridge University Press.

Bates, Elizabeth, Camaioni, Luigia and Volterra, Virginia. 1975. "The acquisition of performatives prior to speech." *Merrill-Palmer Quarterly* 21: 205–26.

Bennett, Maxwell R. and Hacker, Peter M.S. 2003. *Philosophical Foundations of Neuroscience*. Oxford: Blackwell.

Boesch, Christophe 2007. "What makes us human (*Homo sapiens*)? The challenge of cognitive cross species comparison." *Journal of Comparative Psychology* 121: 227–240

Buller, David J. 2005. *Adapting Minds: Evolutionary Psychology and the Persistent Quest for Human Nature*. Cambridge, MA: MIT Press.

D'Entremont, Barbara and Seamens, Elizabeth. 2007. "Do infants need social cognition to act socially? An alternative look at infant pointing." *Child Development* 78: 723–728.

Glock, Hans-Johann. 2009. "Can animals act for reasons?" *Inquiry* 52: 232–254.

Gómez, Juan-Carlos. 2007. "Pointing behaviours in apes and human infants: A balanced interpretation." *Child Development* 78: 729–734.

Gottlieb, Gilbert. 2002. *Individual Development and Evolution: The Genesis of Novel Behaviour*. Mahwah, NJ: Erlbaum.

Hark, Michel ter. 2006. "Wittgenstein, pretend play and the transferred use of language." *Journal for the Theory of Social Behaviour* 36: 299–318.

Jablonka, Eva and Raz, Gal. 2009. "Transgenerational epigenetic inheritance: Prevalence, mechanisms, and implications for the study of heredity and evolution." *The Quarterly Review of Biology* 84: 131–176.

Johnson, Christine M. 2001. "Distributed primate cognition: A review." *Animal Cognition* 4: 167–83.

Leavens, David A. and Racine, Timothy P. 2009. "Joint attention in apes and humans: Are humans unique?" *Journal of Consciousness Studies* 16: 240–267.

Leavens, David A., Hopkins, William D. and Bard, Kim A. 2008. "The heterochronic origins of explicit reference." In *The Shared Mind: Perspectives on intersubjectivity*, Jordan Zlatev, Timothy P. Racine, Chris Sinha and Esa Itkonen (eds), 187–214. Amsterdam/Philadelphia: John Benjamins Publishing Company.

Leavens, David A., Racine, Timothy P. and Hopkins, William D. 2009. "The ontogeny and phylogeny of non-verbal deixis." In *The Cradle of Language. Volume 1: Multidisciplinary Perspectives*, Rudie Botha and Chris Knight (eds), 142–165. Oxford: Oxford University Press.

Leudar, Ivan and Costall, Alan. 2004. "On the persistence of the 'problem of other minds' in Psychology: Chomsky, Grice and theory of mind." *Theory & Psychology* 14: 601–621.

Leslie, Alan M. 1991. "The theory of mind impairment in autism: Evidence for a modular mechanism of development?" In *Natural Theories of Mind*, Andrew Whiten (ed.), 19–46. Oxford: Blackwell.

Lewontin, Richard C. 2000. *The Triple Helix: Gene, Organism, and Environment*. Cambridge, MA: Harvard University Press.

Lickliter, Robert. 2008. "The growth of developmental thought: Implications for a new evolutionary psychology." *New Ideas in Psychology* 26: 353–369.

Liszkowski, Ulf, Carpenter, Malinda, Striano, Tricia and Tomasello, Michael. 2006. "Twelve- and 18-month olds point to provide information for others." *Journal of Cognition and Development* 7: 173–187.

Machado, Armando and Silva, Francisco J. 2007. "Toward a richer view of the scientific method: The role of conceptual analysis." *American Psychologist* 7: 671–681.

Moore, Chris and Corkum, Valerie. 1994. "Social understanding at the end of the first year of life." *Developmental Review* 14: 349–372.

Penn, Derek C. and Povinelli, Daniel J. 2007. "On the lack of evidence that chimpanzees possess anything remotely resembling a 'theory of mind.'" *Philosophical Transactions of the Royal Society* 362: 731–744.

Pika, Simone. 2008. "Gestures of apes and pre-linguistic human children: Similar or different?" *First Language* 28: 116–40.

Pika, Simone and Zuberbühler, Klaus. 2008. "Social games between bonobos and humans: Evidence for shared intentionality?" *American Journal of Primatology* 70: 207–210.

Racine, Timothy P. in press. "Getting beyond rich and lean views of joint attention." In *Joint attention: Developments in philosophy of mind, developmental and comparative psychology, and cognitive science*, Axel Seemann (ed.). Cambridge, MA: MIT Press.

Racine, Timothy P. and Carpendale, Jeremy I. M. 2007. "The role of shared practice in joint attention." *British Journal of Developmental Psychology* 25: 3–25.

Racine, Timothy P. and Müller, Ulrich. 2009. "The contemporary relevance of Wittgenstein: Reflections and directions." *New Ideas in Psychology* 27: 107–117.

Racine, Timothy P., Wereha, Tyler J. and Leavens, David A. (in press). "To what extent nonhuman primates are intersubjective and why." In *Moving ourselves, moving others: Motion and emotion in intersubjectivity, consciousness and language*, Ad Foolen, Jordan Zlatev, Ulrike Lüdtke and Timothy P. Racine (eds). Amsterdam/Philadelphia: John Benjamins Publishing Company.

Racine, Timothy P., Leavens, David A., Susswein, Noah and Wereha, Tyler J. 2008. "Conceptual and methodological issues in the investigation of primate intersubjectivity." In *Enacting Intersubjectivity: A Cognitive and Social Perspective to the Study of Interactions*, Francesca Morganti, Antonia Carassa and Giuseppe Riva (eds), 65–79. Amsterdam: IOS Press.

Savage-Rumbaugh, E. Sue, Shanker, Stuart G. and Taylor, Talbot J. 1998. *Apes, Language, and the Human Mind*. Oxford: Oxford University Press.

Searle, John R. 1990. "Collective intentions and actions." In *Intentions in Communication*, Philip R. Cohen, Jerry Morgan and Martha E. Pollack (eds.), 401–415. Cambridge, MA: MIT Press.

Seemann, Axel (ed.). in press. *Joint Attention: Developments in Philosophy of Mind, Developmental and Comparative Psychology, and Cognitive Science.* Cambridge, MA: MIT Press.

Sharrock, Wes and Coulter, Jeff. 2004. "ToM: A critical commentary." *Theory & Psychology* 14: 579–600.

Southgate, Victoria, van Maanen, Catherine and Csibra, Gergely. 2007. "Infant pointing: Communication to cooperate or communication to learn?" *Child Development* 78: 735–740.

Susswein, Noah and Racine, Timothy P. 2008. "Sharing mental states: Causal and definitional issues in intersubjectivity." In *The Shared Mind: Perspectives on Intersubjectivity*, Jordan Zlatev, Timothy P. Racine, Chris Sinha and Esa Itkonen (eds), 141–162. Amsterdam/Philadelphia: John Benjamins Publishing Company.

Tomasello, Michael. 1999. *The Cultural Origins of Human Cognition.* Cambridge, MA: Harvard University Press.

Tomasello, Michael. 2006. "Why don't Apes Point?" *In Roots of Human Sociality: Culture, Cognition and Interaction,* Nicholas J. Enfield and Stephen C. Levinson (eds), 506–524. Oxford: Berg.

Tomasello, Michael. 2008. *Origins of Human Communication.* Cambridge, MA: MIT Press.

Tomasello, Michael. 2009. *Why We Cooperate.* Cambridge, MA: MIT Press.

Tomasello, Michael and Carpenter, Malinda. 2005. "The emergence of social cognition in three young chimpanzees." *Monographs of the Society for Research in Child Development*, 70 (Serial no. 279).

Tomasello, Michael and Carpenter, Malinda. 2007. "Shared intentionality." *Developmental Science* 10: 121–125.

Tomasello, Michael, Carpenter, Malinda and Liszkowski, Ulf. 2007. "A new look at infant pointing." *Child Development* 78: 705–722.

Tomasello, Michael, Carpenter, Malinda, Call, Josep, Behne, Tanya and Moll, Henrike. 2005. "Understanding and sharing intentions: The origins of cultural cognition." *Behavioural and Brain Sciences* 28: 675–735.

Warneken, Felix, Chen, Frances and Tomasello, Michael. 2006. "Cooperative activities in young children and chimpanzees." *Child Development* 77: 640–663.

Wereha, Tyler J. and Racine, Timothy P. in press. "Evolution, development, and human social cognition." *Review of Philosophy and Psychology.*

Wittgenstein, Ludwig. 2009. *Philosophical Investigations,* Peter M. S. Hacker and Joachim Schulte (eds). Oxford: Wiley-Blackwell.

Pointing

Contexts and instrumentality

David A. Leavens
University of Sussex

Although long heralded as a human species-unique gesture, POINTING has now been demonstrated in numerous species of non-human animals. Many contemporary researchers argue that POINTING for instrumental ends marks a different kind of psychological process from POINTING to share attention as an end in itself. Thus, a large body of contemporary theory is built on presumptions about the hypothetical motivations underlying POINTING. I will briefly outline some of the contexts and motivations in which humans POINT, and argue that virtually all human POINTING can be interpreted in instrumental terms. If this is correct, then instrumentality, *per se*, cannot illuminate the evolutionary origins of joint attention.

There is a contemporary debate over the evolutionary origins of the capacity to share attention, jointly, with another organism, a capacity that seems to be foundational for learning words (e.g. Baldwin 1995). Representatives of one camp argue that this is a human-unique ability, reflecting our species' independent evolutionarily adaptive history since our split with the chimpanzee/bonobo lineage, about six million years ago (e.g. Butterworth 2003; Povinelli, Bering & Giambrone 2003; Tomasello Carpenter & Liszkowski 2007). Others argue, in contrast, that joint attention is, to at least a very large degree, shared by humans and their nearest living relatives, the great apes (e.g. Bard & Leavens 2009; Leavens & Bard 2011; Leavens & Racine 2009; Leavens, Hopkins & Bard 2005a, 2008; Leavens, Racine & Hopkins 2009; Racine, Leavens, Susswein & Wereha 2008; Savage-Rumbaugh, Rumbaugh & MacDonald 1985). POINTING to a particular location is a quintessential example of joint attention. POINTING with the index finger has been described as both a "universal" and a biologically given species-specific gesture in humans (e.g. Butterworth 2003; Eibl-Eibesfeldt 1989). However, recent evidence demonstrates that index-finger POINTING is neither universal nor species-specific to humans (see reviews by Bard & Leavens 2009; Leavens 2004; Leavens & Bard

2011; Leavens & Hopkins 1999; Leavens & Racine 2009; Leavens et al. 2009, on great apes; and Wilkins 2003, on humans). Moreover, great apes and humans POINT not only with their index fingers, but also with other single digits (Leavens, Hopkins & Bard 1996), with all fingers extended (Call & Tomasello 1994; Leavens et al. 1996; Wilkins 2003), with their lips (Enfield 2001; Kellogg & Kellogg 1933; Wilkins 2003), and even with their feet (Woodruff & Premack 1979). Therefore, there is neither cross-cultural nor cross-species validity in the older view of index-finger POINTING as a biologically given human species-specific gesture. Indeed, POINTING has been reported in captive monkeys (Blaschke & Ettlinger 1987; Hess, Novak & Povinelli 1993; Kumashiro, Ishibashi, Itakura & Iriki 2002; Mitchell & Anderson 1997). POINTING with the long axis of the body has been reported in dolphins (Xitco, Gory & Kuczaq 2001, 2004) and Australian magpies (Kaplan 2011).

Among humans, POINTING is a culturally shaped gesture, in structural terms: for some people the canonical POINTING gesture is LIP-POINTING (e.g. Enfield 2001; Wilkins 2003), whereas for others it is POINTING with the index finger extended. In functional terms, some kind of POINTING gesture does seem to be a human universal – as far as I know, no human population has ever been reported to lack deictic gestures in their repertoires.

Among great apes, POINTING is only rarely reported in wild populations (e.g. Hobaiter 2010; Veà & Sabater-Pi 1998; and see Pika & Mitani 2006, for evidence of referential scratching in wild chimpanzees), but it is frequently observed in institutionalized populations, and ubiquitous among home-raised and language-trained apes (henceforth, "cross-fostered" apes; see mini-review by Leavens, Bard & Hopkins 2010). Indeed, I am unaware of a single cross-fostered ape who does not POINT. Therefore, it is clear that apes' POINTING is profoundly influenced by their rearing contexts.

A central pillar of the view that humans have unique psychological motivations for POINTING is the oft-repeated claim that great apes POINT only for imperative purposes; that is, only to instrumentally manipulate somebody into acting on the world for them – for example, POINTING through cage mesh to desirable food for the apparent purposes of requesting delivery of the food (Tomasello et al. 2007). In contrast, I argue here that POINTING has instrumental consequences across a wide range of apparent motivations. We may POINT to identify an object we want delivered (for example, the mashed potatoes on the dinner table), to command somebody to look at something, to choose someone for a team, to inform a dinner partner of a piece of bread stuck to his beard, or the location of an object for which they are looking; in all of these cases, the POINTING is instrumental. POINTING is a means to diverse ends.

1. Pointing towards social ends

POINTING by human infants towards social ends was described by Bates and her colleagues (1975). They defined two kinds of instrumental communication by human infants: *proto-imperative* and *proto-declarative*. Proto-imperative communication included POINTING to obtain objects, as when a child POINTS to an object to request its delivery, in other words, the instrumental use of a social agent (the means) to obtain an object (the end). They defined proto-declarative communication as the instrumental use of an object (the means) to obtain a social response (the end). Examples include the showing of an object to elicit a reaction and, similarly, POINTING to an object to achieve a similar (social) end. Naming behavior, such as POINTING to a dog, while uttering, "Doggie!" is quintessentially proto-declarative. In this example, the child uses the dog and its label to elicit a reaction from their social partner.

Although the use of the term "proto-imperative" has changed little in the intervening decades, the use of the term "proto-declarative" has undergone a profound equivocation in meaning. In contemporary uses of the term, it has come to signify (a) a human-unique capacity for the representation of mental states in others (e.g. Baron-Cohen 1995) or (b) a human-unique motivation for sharing attention (e.g. Tomasello et al. 2007). There is now a widespread belief in psychology that proto-declarative communication by children as young as twelve months of age implies their "understanding of the mental states of others" (Tomasello et al. 2007: 715).

Numerous critiques of these rich interpretations of proto-declarative communication in human infants exist in the literature (e.g. Moore & Corkum 1994; Leavens 2004, in press; Leavens et al. 2009; Racine & Carpendale 2007). Some of these critiques are grounded in the view that to display intentional gestures for non-instrumental ends is logically self-contradictory (see Liszkowski, Carpenter, Henning, Striano & Tomasello 2004 for an experimental demonstration of the goal-directedness of proto-declarative POINTING in human one-year-olds). Here, however, I wish to emphasize that the physical circumstances of proto-imperative and proto-declarative POINTING differ in many dimensions both across and within laboratory studies. In other words, irrespective of the hypothetical, invisible psychological processes that some researchers argue differentiate proto-imperative from proto-declarative POINTING, the fact of the matter is that researchers elicit proto-imperative POINTING from human infants using different kinds of experimental setups, different kinds of targets, and different experimenter behaviors (before, during, and after the infants' POINTS), compared with the contexts in which proto-declarative POINTING is elicited.

This is particularly clear in a study designed to elicit both imperative and declarative POINTING by Franco and Butterworth (1996). In this classic study, imperative POINTS were elicited by the presentation, just out of reach, of small objects (e.g. toy phones) that fit easily into the hand, whereas declarative POINTING was elicited by the presentation, at a large distance, of large, planometric figures, that could not fit easily into the hand. Thus, both the distances and specific features of the eliciting stimuli varied systematically between conditions. And this is generally true for all experimental studies of POINTING – it is technically difficult to define any single POINTING context that will elicit POINTS of different alleged motivations. This means that the different motivations for POINTING are irremediably entangled with the physical circumstances of their elicitation. The implication of this fact for understanding infants' communicative motivations is profound: it is not possible to objectively define a proto-declarative motivation that is something separate from a proto-declarative procedure or experimental context. Thus, when young human children POINT to a desirable, but unreachable object that fits easily within the hand, researchers argue that the underlying psychological process is different from when they POINT to a large, colorful poster on a distant wall, but note that the physical circumstances are also radically different. A proto-declarative POINT is not, therefore, something that is defined only by the invisible hypothetical mental processes of the signalling child, but also by its eliciting contexts; a proto-declarative POINT implies a proto-declarative context.

With respect to the putative human species-specificity of proto-declarative POINTING (Baron-Cohen 1995; Povinelli et al. 2003; Tomasello et al. 2007), children enter laboratory studies with uncontrolled developmental histories in which they have experienced an unknown number of proto-declarative contexts. Hence, to some degree, children bring task-relevant experience into studies of their POINTING behavior. The central problem in comparing human children with representatives of our nearest living relatives, the great apes, is that relatively few apes will have had anything remotely close to the same number of exposures to proto-declarative contexts, compared with human children (Leavens 2004; Leavens et al. 2009; Racine et al. 2008).

There are, thus, two different levels at which claims for human uniqueness in proto-declarative signaling are problematic: first, in their portrayal of these signals as "non-instrumental" and, secondly, in the widespread failure to compare humans and apes of similar pre-experimental backgrounds. It should be obvious that an orphaned ape raised in a cage without ever having played "peek-a-boo" or been read to from a picture book has not been exposed to the same task-relevant experience as a typical child in historically unprecedented, geographically limited, Western, post-industrial families (Bard & Leavens 2009; Leavens & Bard 2011). To summarize, proto-declarative signaling is displayed in particular contexts by particular organisms who have particular levels of pre-experimental exposures to proto-declarative contexts.

Recently, Liszkowski and his colleagues (2006) claimed that there is a third motivation for POINTING displayed by human infants: to inform others. For example, in their study, human one-year-olds POINTED to items for which the experimenter was apparently looking. Although there is some debate about the validity of this kind of signaling as something different from proto-declarative signaling, nevertheless, there is some empirical support for the idea. For example, Cochet and Vauclair (2010) measured the hand used and a number of behavioral accompaniments to POINTING (including vocalizations and gaze alternation) in what they termed imperative, declarative-expressive, and declarative-informative contexts. They found that in terms of gaze-alternating behavior, declarative-informative POINTS were differentiated from declarative-expressive POINTS. Specifically, the children alternated their gaze between a referent and an experimenter more often in declarative-informative contexts than in declarative-expressive contexts. Still, as for proto-declaratives, it remains the case that informative POINTING implies something about the physical contexts in which it is elicited and the pre-experimental histories of experiences with those informative contexts. In this study, the eliciting stimuli, distances between the children and these stimuli, and the behavior of the experimenter all differed between the different experimental contexts. Therefore, as for proto-declarative signaling, informative POINTING is defined, in part, by physical features outside the brain of the signaling organism. Moreover, informative POINTING is well-described in apes (see Call 2011; Leavens & Bard 2011; Pedersen, Segerdahl & Fields 2009), so it is far from a uniquely human behavior.

As Racine and his colleagues have argued, whether one or another kind of POINT is a more appropriate label depends crucially on the specific parameters of the contexts in which the gesture is displayed (Racine et al. 2008). As they put it, "we say that the motive is declarative (or imperative) *because of the situation in which it is embedded*" (2008: 70, emphasis in original).[1]

2. Why do people point?

In Table 1, I list 18 commonplace reasons that adult humans POINT. This list is not meant to be comprehensive, merely illustrative. I have tried to capture, in ordinary

1. In contemporary writing, the prefix, "proto-" is often dropped, and researchers write as though a "proto-declarative" motivation is the same thing as a "declarative" motivation. In fact, this is an unwarranted assumption (e.g. Leavens & Hopkins 1998), but this issue is beyond the scope of this chapter. For present purposes, I will simply drop the prefix "proto-" henceforth.

Table 1. Common human uses of non-linguistic POINTING (excluding use of POINTS in sign languages)

POINTING episodes	Dichotomous (Bates et al. 1975)	Trichotomous (Liszkowski et al. 2006)	Skinner (1957)	Searle (1975)	Observed in Apes?
1. POINTING for self	?	?	Mand/Tact/ Intraverbal	?	Yes[a]
2. To request or demand an action on an object	Imperative	Imperative	Mand	Directive	Yes[b]
3. To draw attention to self or part of self	Imperative/ Declarative	Imperative/Declarative/ Informative	Mand	Directive	Yes[c]
4. To draw attention to a broken apparatus	Imperative/ Declarative	Imperative/Declarative/ Informative	Mand/Tact	Directive/ Representative/ Expressive	Yes[d]
5. To draw attention to misplaced or hidden entities	Imperative/ Declarative	Imperative/Declarative/ Informative	Mand/Tact/ Intraverbal	Directive/ Representative	Yes[e]
6. To request repositioning of social partner	Imperative	Imperative	Mand/Intraverbal	Directive	Yes[f]
7. To warn of impending threat	Imperative/ Declarative	Informative	Mand/Tact	Directive	Yes[g]
8. To misdirect or deceive	Imperative/ Declarative?	Imperative/Declarative/ Informative?	Tact?	Representative	Yes[h]
9. To answer a request for directions	Declarative?	Informative	Intraverbal	Representative?	Yes[i]
10. To request information about something	Imperative	Imperative	Mand	Directive	Yes[j]

(Continued)

Table 1. (Continued)

11. Whilst naming an entity	Imperative/ Declarative	Imperative/Declarative/ Informative	Tact	Representative	Yes[k]
12. To draw attention to an entity without naming it	Imperative/ Declarative	Imperative/Declarative/ Informative	Mand/Intraverbal/ Tact	Representative	Yes[l]
13. To insult	Declarative?	Declarative?	Mand/Tact	Directive/ Expressive	No
14. To threaten physical harm	Declarative	Declarative	Mand	Commissive	No
15. To curse (spiritual harm)	Declarative	Declarative	?	Declaration/ Expressive	No
16. To glorify or exalt	Declarative	Declarative	Intraverbal	Expressive	No
17. To thank	Declarative	Declarative	Intraverbal	Expressive	No
18. To acknowledge	Declarative	Declarative	Intraverbal	Expressive	No

Notes: Representative sources for the observations in great apes. [a]Boysen et al. 1995. [b]Call & Tomasello 1994; Leavens & Hopkins 1998; Leavens et al. 1996, 2004b, 2005b; Woodruff & Premack 1979. [c]Hoyt 1941; Tomasello, George, Kruger, Farrar & Evans 1985; Yerkes 1943. [d]Leavens et al. 1996; Rumbaugh 1977. [e]Call & Tomasello 1994; Leavens et al. 2004b; Liszkowski et al. 2009; Menzel 1999. [f]Savage-Rumbaugh, Wilkerson & Bakeman 1977; Tanner & Byrne 1996. [g]Veà & Sabater-Pi 1998. [h]Woodruff & Premack 1979. [i]Menzel 1999; Savage-Rumbaugh et al. 1985, 1986. [j]Savage-Rumbaugh, Shanker & Taylor 1998. [k]Menzel 1999. [l]Carpenter, Tomasello & Savage-Rumbaugh 1995; Furness 1916; Kellogg & Kellogg 1933; Leavens & Hopkins 1998; Leavens et al. 1996, 2004b, 2005b; Witmer 1909.

English, a few prosaic examples of POINTING from common experience. Mapping dichotomous (imperative, declarative) or trichotomous (imperative, declarative and informative) motivations on these commonplace examples is a truly Procrustean exercise. As perusal of Table 1 makes clear, it is often entirely ambiguous whether any particular POINT is better characterized as an "imperative" or a "declarative" (or "informative"). For example, a child may POINT to its own arm (a) to seek alleviation from pain (an imperative POINT), to (b) to POINT out a contusion or bruise (a declarative POINT), or (c) in response to the query, "Where does it hurt?" (an informative POINT). Under some circumstances, we might imagine, realistically, that a child has all three motivations either simultaneously or near-simultaneously with the emission of a single POINTING gesture. Here, although the immediate context is the same, the interpretation of these signals can differ in accordance with the specific learning histories not only of the child, but of the interpreter, as well. As discussed above, most of the gestures listed in Table 1 are displayed in specific contexts between specific individuals who are in some kind of relationship with each other, with relationship histories of various durations. Nothing in the structure of a POINTING gesture, itself, strictly determines its function (see, for example, Cochet & Vauclair 2010; Liebal, Behne, Carpenter & Tomasello 2009; Tomasello et al. 2007). It is, therefore, debatable whether imperative, declarative, and informative motivations can ever be treated, scientifically, as attributes of the child's hypothetical, invisible psychological processes, separately from that child's immediate physical context and pre-experimental learning history. To summarize, labels such as "imperative," "declarative," and "informative" specify physical parameters of the contexts in which children and apes POINT and the interpretive biases of observers; they do not characterize aspects of these organisms' invisible psychological processes, because these hypothetical processes are not available for direct scientific scrutiny (Leavens et al. 2004a). Hence, these defining attributes of the different motivations for POINTING are not, themselves, direct aspects of the child's invisible psychological processes; rather, they are objective characteristics of the universe, open to public scrutiny (Leavens & Bard 2011; Racine et al. 2008). POINTING gestures are not inherently imperative, declarative, or informative, they are displayed in imperative, declarative, or informative contexts, and these contexts are physically different from other contexts in multitudinous ways. In fact, however, there are many more alleged motivations to communicate than the three discussed, so far.

3. Skinner and Searle on communication

Many people have classified communicative signals in terms of contexts beyond those we have considered, above. For purposes of comparison, I have applied two previous, systematic approaches to communicative contexts to describe these motivational aspects of POINTING, including Skinner's (1957) scheme, and Searle's (1975) list of speech motivations. Skinner (1957) described several categories of verbal behavior (which, for him, was not limited to linguistic behavior) characterized by different contingency structures. A *mand* is followed by a conventionalized or expected response and controlled by conditions of "deprivation or aversive stimulation" (Skinner 1957:36). According to Skinner, mands specify their reinforcer: to take his example, "Please pass the bread." The command "Look!" is a mand, and this clarifies the nature of the contemporary dispute over declarative and imperative POINTING: some researchers think that declarative POINTING is a mand (a command to look), at least in the usual laboratory contexts in which it is elicited (e.g. Moore & Corkum 1994; Leavens et al. 2005b), whereas most workers in this area consider it to be a *tact* (an expression about the state of the world). To oversimplify, a tact is a kind of labelling response, which is under the control of generalized reinforcement; correct labelling in interactive contexts leads to rewarding responses from our community (for example, when a child POINTS to a dog while saying, "Doggie!"), but in contrast to the mand, the reinforcer is not specified by the tact. Skinner offers an heuristic to distinguish mands and tacts:

> [r]oughly speaking, the mand permits the listener to infer something about the condition of the speaker regardless of the external circumstances, while the tact permits him to infer something about the circumstances regardless of the condition of the speaker. (p. 83)

Finally, Skinner described an *intraverbal* response, in which the eliciting stimulus is the verbal behavior of an interlocutor. For example, a child might POINT to a dog when an interlocutor asks, "Where is the doggie?" Thus, the intraverbal is a response to an explicit query, as well as a conventionalized response to implicit demands for a response. Saying "thank you" after somebody passes the bread at the dinner table is an intraverbal response – it is a conventionalized response to the receipt of a favor.[2]

As we apply Skinner's terminology to the various kinds of POINTS in Table 1, we find that there is a kind of intuitive fit between mands and imperatives, but, overall, it is often ambiguous whether any particular kind of POINTING gesture is better characterized as a mand, a tact, or an intraverbal. For example, whether POINTING to a broken apparatus is a mand, a tact, or an intraverbal depends upon contextual elements that are not specified in the bare description. So when Clint, a young adult

2. Skinner described several other verbal responses, such as *echoics* and *textuals*, which need not concern us, here.

Figure 1. Clint POINTS repeatedly at malfunctioning computer. The joystick has frozen, leaving a cursor unresponsive to Clint's manipulations. In Panel A, Clint quietly awaits the return of the experimenter, holding on to the cage mesh with his left hand, without gesturing. In Panel B, as the experimenter enters (offscreen), Clint rotates his hand and extends his fingers in a POINT to the computer monitor. In Panel C, as the experimenter leans in to peer at the monitor screen (arrow), Clint relaxes his fingers. In Panel D, Clint has retracted his hand and POINTS again. In Panel E, Clint has retracted his hand and POINTED for a third time. In Panel F, Clint stops gesturing as the experimenter begins to remove the joystick (arrow). The entire communicative sequence lasts 14 seconds. Images captured from digital files originally recorded on VHS videotapes. This video archive formed the basis for the analysis presented in Leavens et al. 1996.

chimpanzee, POINTED to a broken computer apparatus (Leavens et al. 1996), was he demanding that I fix it (a mand) or expressing his dismay at the interruption of his activity (a tact), or both (see Figure 1)? When someone POINTS to re-position a social partner, it might be a mand or an intraverbal, depending upon whether the receiver had signaled his or her desire for direction. As noted above, an organism might draw a receiver's attention to a hidden or misplaced object with a POINT that can be classified as a mand, a tact, or an intraverbal, depending upon whether the signaler also communicates its apparent desire for the receiver to retrieve the object (mand), its apparent desire for the receiver to regard the object (tact), or is simply responding to the verbal or non-verbal search behavior of the receiver (intraverbal). Thus, to use Skinner's classification scheme to classify POINTING, one has to know what the proximal antecedents are of a POINTING gesture's use, what are the behavioral accompaniments, as well as something about the history of the signaling organism's exposure to contexts of manding, tacting, and intraverballing.

Searle's (1975) approach to performatives is also contextually based. However, in contrast to the approach taken by Skinner (1957), Searle discriminated among the different motivation categories in relation to the propositional content of an utterance. Hence, his *representative* – which corresponds to the tact – is a commitment to the truth value of the associated proposition (e.g. "Doggie!"). Searle's *directive* maps fairly easily onto both terms: (proto-)imperative and mand. Searle does introduce, however, some additional distinctions that capture certain uses of POINTING gestures that are not as well represented by the taxonomies we have already discussed. For example, as this chapter is written, it is fairly common to witness contemporary athletes POINTING skyward in explicit thanks for divine intervention during an athletic competition. In Searle's scheme, these POINTS are called *expressives*; these performatives serve to express a "psychological state" (1975: 356). Expressives can also be used to glorify or exalt; POINTING skywards by athletes might constitute exaltations, rather than thanks. One example of an exaltatory POINT from my experience was displayed by the guitarist, Carlos Santana, at the John Coltrane Festival, in the late 1980s. After a scintillating performance of his Grammy-winning instrumental, Blues for Salvador, the audience erupted into a standing ovation. Mr. Santana POINTED repeatedly at an immense photograph of John Coltrane, in an apparent attempt to deflect the exaltation away from himself and towards the famed jazz saxophonist. Expressives can also be used to insult; for example, Eibl-Eibesfeldt (1989) describes "pointing-duels" between young boys (pp. 481, 485). *Commissives*, in Searle's taxonomy, are "acts whose point is to commit the speaker … to some future course of action" (1975: 356). An example of a commissive POINT occurred during my U.S. Marine Corps training in pugil sticks. This is a kind of bayonet training in which sticks with padding at each end are used in hand-to-hand combat. A large, muscular member of a rival platoon repeatedly POINTED to me in direct threats, commitments to future actions in which he promised to pummel me into submission. A final Searlean category of performative, relevant to the present discussion, is that of the *declaration*. A declaration creates a state of the world in its performance. An example of a linguistic declaration is when a minister or a justice of the peace states, "I now pronounce you, 'man and wife.'" The state of matrimony is achieved by virtue of the speech act. An example of a POINTING declaration is the POINTING-TO-CURSE gesture, which has a widespread distribution. For example, in the *kurdaitcha* ritual of central Australia, a shaman POINTS a specially prepared bone or other sharp object at the accursed, bringing the curse into being. Searle's taxonomy fares well in comparison to the other schemes considered, here, but it is, again, sometimes ambiguous whether one category or another is best applied, in the absence of a richer representation of the episode in which organisms POINT.

Thus, the important thing to notice about Table 1 is that the same sort of gesture can have multiple labels, depending on context, and that there is relatively little agreement between systems for many kinds of POINTING. When a child POINTS for him- or herself, for example, in an asocial manner, or "private pointing" (Delgado, Gómez & Sarrià 2009; Rodriguez 2009), is this a kind of self-declarative or self-imperative? In Rodriguez's (2009) analysis, people often POINT privately in a self-interrogative mode. People also POINT, privately, whilst counting (Fuson 1988, cited in Boysen, Berntson, Shreyer & Hannan 1995). So, private POINTING has varied ends and seems to be used as an aid in cognitive or perceptual tasks typical of western upbringing (see Boysen et al. 1995 for evidence of private POINTING during counting by a chimpanzee). Private POINTING seems to have, thus, a variety of different functions, none of which are well-captured by a dichotomous or trichotomous classification, or by Searle's (1975) system (see Rodriguez 2009, for an extended analysis). In Skinner's (1957) terminology, private POINTING can be construed as a kind of private speech or an accompaniment to private speech, so is interpretable within his scheme, but whether it is a mand, tact, or intraverbal is completely dependent on the circumstances. For example, if a person thought to themselves, "Where is that book by Hoyt?", then scanned some bookshelves, POINTING to it while thinking, "Oh, there it is!", this would be an example of an intraverbal. Alternatively, if one were to POINT while counting, this might be a kind of tact. Thus, even with a scheme that can explicitly handle private POINTING, the gestures need at least three labels (e.g. "self-pointing" as a mand, as tact, or as an intraverbal) to accommodate the varied circumstances in which it is displayed.

4. Conclusions

The central conclusion of this brief sketch is that all categories of POINT listed in Table 1 are instrumental, with the possible exception of private POINTING (although it may be that private POINTING serves as an aid to ongoing cognition, which would therefore make it an instrumental act – see, esp. Goldin-Meadow 2009, and references therein, and Rodriguez 2009). In other words, these gestures are displayed to effect changes in the world. These changes are, in the overwhelmingly majority of instances, measurable in objective terms. It is, in most ordinary circumstances, a logical contradiction to assert that any signal is displayed with a particular motivation, but that it is "non-instrumental." Exceptions to this rule might involve the kinds of gesticulations displayed by people suffering from mental illness, but, barring these pathological circumstances, POINTING serves, in humans and apes, a relatively wide variety of ends.

Secondly, these different categories of POINTING are not distinguished exclusively by the invisible psychological processes that an observer might attribute to the signaler, but also by patterns of history and circumstance that lie entirely outside the crania of the POINTING individuals; it is only these external events that we can directly measure or control. This has immense significance for understanding the acquisition of POINTING: if different POINTING motivations are discriminable in physical terms, then organisms do not need to infer the invisible contents of others' "minds" to acquire and correctly use these gestures (see Leavens & Racine 2009; Leavens et al. 2004a; Leudar & Costall 2004). Questions of group differences in the propensities of representatives of different groups (either within or between species) to POINT in various ways will, therefore, necessarily require that researchers explicitly control their experience with objectively defined physical circumstances corresponding to the kinds of circumstances in which organisms, for example, POINT "declaratively" or "informatively." The fact of the matter is that human babies, typically, experience many more episodes of engagement in "declarative" and "informative" circumstances than do any non-human animals, except in the most extraordinary of captive circumstances (e.g. Leavens & Bard 2011; Leavens et al. 2008; Racine et al. 2008). Experimental protocols designed to elicit one kind of POINTING differ in physical terms from experimental protocols to elicit other kinds of POINTING; children and apes will have varying levels of experience with these specific contexts (Leavens & Bard 2011).

Finally, it is worth noting that, of the 18 commonplace different kinds of POINTS listed in Table 1, only the last six have yet to be observed in our nearest living relatives, the great apes. There are numerous false claims to the effect that great apes only POINT imperatively (e.g. Povinelli et al. 2003; Tomasello et al. 2007).

Perusal of the references provided in Table 1 should dispel this curiously popular, yet wholly inaccurate myth (see Leavens in press; Leavens & Bard 2011; Leavens & Racine 2009).

Acknowledgements

I am grateful to many researchers for fruitful discussions about some of the issues raised in this chapter, including especially my collaborators, Dr. William D. Hopkins, Prof. Kim A. Bard, and Dr. Timothy P. Racine. Thanks to Prof. Steve Torrance for inspiring this line of inquiry.

References

Baldwin, Dare A. 1995. "Understanding the link between joint attention and language." In *Joint attention: Its origins and role in development,* Chris Moore and Philip J. Dunham (eds), 131–158. Hillsdale, NJ: Lawrence Erlbaum.

Bard, Kim A. and Leavens, David A. 2009. "Socio-emotional factors in the development of joint attention in human and ape infants." In *Learning from animals? Examining the nature of human uniqueness*, Louise Röska Hardy and Eva M Neumann Held (eds), 89–104. London: Psychology Press.

Baron-Cohen, Simon. 1995. *Mindblindness: An essay on autism and theory of mind.* Cambridge, Mass.: MIT Press.

Bates, Elizabeth, Camaioni, Luigia and Volterra, Virginia. 1975. "Performatives prior to speech." *Merrill-Palmer Quarterly* 21: 205–26.

Blaschke, M. and Ettlinger, George. 1987. "Pointing as an act of social communication by monkeys." *Animal Behaviour* 35: 1520–1523.

Boysen, Sarah T., Berntson, Gary G., Shreyer, Traci A. and Hannan, Michelle B. 1995. "Indicating acts during counting by a chimpanzee (*Pan troglodytes*)." *Journal of Comparative Psychology* 109: 47–51.

Butterworth, George. 2003. "Pointing is the royal road to language for babies." In *Pointing: Where language, culture, and cognition meet,* Sotaro Kita (ed.), 9–33. Mahwah, NJ: Erlbaum.

Call, Josep. 2011. "How artificial communication affects the communication and cognition of the great apes." *Mind & Language* 26: 1–20.

Call, Josep and Tomasello, Michael. 1994. "Production and comprehension of referential pointing by orangutans (*Pongo pygmaeus*)." *Journal of Comparative Psychology* 108: 307–317.

Carpenter, Malinda, Tomasello, Micheal and Savage-Rumbaugh, Sue E. 1995. "Joint attention and imitative learning in children, chimpanzees, and enculturated chimpanzees." *Social Development* 4: 217–237.

Cartmill, Erica A. and Byrne, Richard W. 2007. "Orangutans modify their gestural signaling according to their audience's comprehension." *Current Biology* 17: 1345–1348.

Cochet, Hélène and Vauclair, Jacques. 2010. "Pointing gestures produced by toddlers from 15 to 30 months: Different functions, hand shapes and laterality patterns." *Infant Behavior and Development* 33: 431–441.

Delgado, Begoña, Gómez, Juan-Carlos and Sarriá, Encarnación. 2009. "Private pointing and private speech: Developing parallelisms." In *Private speech, executive functioning, and the*

development of verbal self-regulation, Adam Winsler, Charles Fernyhough, and Ignacio Montero (eds), 153–162. Cambridge: Cambridge University Press.
Eibl-Eisefeldt, Irenäus. 1989. *Human ethology*. New York: Aldin de Gruyter.
Enfield, N. J. 2001. "'Lip-pointing': A discussion of form and function with reference to data from Laos." *Gesture* 1: 185–212.
Franco, Fabia and Butterworth, George. 1996. "Pointing and social awareness: Declaring and requesting in the second year." *Journal of Child Language* 23: 307–336.
Goldin-Meadow, Susan. 2009. "How gesture promotes learning throughout childhood." *Child Development Perspectives* 3: 106–111.
Hess, Jo, Novak, Melinda A. and Povinelli, Daniel J. 1993. "'Natural pointing' in a rhesus monkey, but no evidence of empathy." *Animal Behaviour* 46: 1023–1025.
Hobaiter, Catherine. 2010. *Gestural communication in wild chimpanzees*. Unpublished doctoral thesis, University of St. Andrews, U.K.
Hoyt, Augusta Maria Daurer. 1941. *Toto and I*. New York: Lippincott.
Kaplan, Gisela. 2011. "Pointing gesture in a bird- merely instrumental or a cognitively complex behavior?" *Current Zoology* 57(4): 453–467.
Kellogg, Winthrop N. and Kellogg, Luella A. 1933. *The ape and the child: A study of early environmental influence upon early behavior*. New York: McGraw-Hill.
Kumashiro, Mari, Ishibashi, Hidetoshi, Itakura, Shoji and Iriki, Atsushi. 2002. "Bidirectional communication between a Japanese monkey and a human through eye gaze and pointing." *Cahiers de Psychologie/Current Psychology of Cognition* 21: 2–32.
Leavens, David A. 2004. "Manual deixis in apes and humans." *Interaction Studies* 5: 387–408.
Leavens, David A. In press. "Joint attention: Twelve Myths." In *Joint attention: New Developments*, Axel Seemann (ed.). Cambridge, MA: MIT Press.
Leavens, David A. and Bard, Kim A. 2011. "Environmental influences on joint attention in great apes: Implications for human cognition." *Journal of Cognitive Education and Psychology* 10: 9–31.
Leavens, David A. and Hopkins, William D. 1998. "Intentional communication by chimpanzees: A cross-sectional study of the use of referential gestures." *Developmental Psychology* 34: 813–822.
Leavens, David A. and Hopkins, William D. 1999. "The whole hand point: The structure and function of pointing from a comparative perspective." *Journal of Comparative Psychology* 113: 417–425.
Leavens, David A. and Racine, Timothy P. 2009. "Joint attention in apes and humans: Are humans unique?" *Journal of Consciousness Studies* 16: 240–267.
Leavens, David A., Bard, Kim A. and Hopkins, William D. 2010. "BIZARRE chimpanzees do not represent 'the chimpanzee'." *Behavioral and Brain Sciences* 33: 100–101.
Leavens, David A., Hopkins, William D. and Bard, Kim A. 1996. "Indexical and referential pointing in chimpanzees (*Pan troglodytes*)." *Journal of Comparative Psychology* 110: 346–353.
Leavens, David A., Hopkins, William D. and Bard, Kim A. (2005a). "Understanding the point of chimpanzee pointing: Epigenesis and ecological validity." *Current Directions in Psychological Science* 14: 185–189.
Leavens, David A., Hopkins, William D. and Bard, Kim A. 2008. "The heterochronic origins of explicit reference." In *The shared mind: Perspectives on intersubjectivity*, Jordan Zlatev, Timothy Racine, Chris Sinha, and Esa Itkonen (eds), 187–241. Amsterdam/Philadelphia: John Benjamins Publishing Company.

Leavens, David A., Hopkins, William D. and Thomas, Roger K. 2004b. "Referential communication by chimpanzees (*Pan troglodytes*)." *Journal of Comparative Psychology* 118:48–57.
Leavens, David A., Hostetter, Autumn B., Wesley, Michael J. and Hopkins, William D. 2004a. "Tactical use of unimodal and bimodal communication by chimpanzees, *Pan troglodytes*." *Animal Behaviour* 67:467–476.
Leavens, David A., Racine, Timothy P. and Hopkins, William D. 2009. "The ontogeny and phylogeny of non-verbal deixis." In *The prehistory of language*, Rudi Botha and Chris Knight (eds), 142–165. Oxford: Oxford University Press.
Leavens, David A., Russell, Jamie L. and Hopkins, William D. 2005b. "Intentionality as measured in the persistence and elaboration of communication by chimpanzees (*Pan troglodytes*)." *Child Development* 76:291–306.
Leudar, Ivan and Costall, Alan. 2004. "On the persistence of the "problem of other minds" in psychology: Chomsky, Grice and theory of mind." *Theory & Psychology* 14:601–624.
Liebal, Kristin, Behne, Tanya, Carpenter, Malinda and Tomasello, Michael. 2009. "Infants use shared experience to interpret pointing gestures." *Developmental Science* 12:264–271.
Liszkowski, Ulf, Carpenter, Malinda, Henning, Anne, Striano, Tricia and Tomasello, Michael. 2004. "Twelve-month-olds point to share attention and interest." *Developmental Science* 7:297–307.
Liszkowski, Ulf, Carpenter, Malinda, Striano, Tricia and Tomasello, Michael. 2006. "12- and 18-month-olds point to provide information for others." *Journal of Cognition and Development* 7:173–87.
Menzel, Charles R. 1999. "Unprompted recall and reporting of hidden objects by a chimpanzee (*Pan troglodytes*) after extended delays." *Journal of Comparative Psychology* 113:426–434.
Mitchell, Robert W. and Anderson, James R. 1997. "Pointing, withholding information, and deception in capuchin monkeys (*Cebus apella*)." *Journal of Comparative Psychology* 111:351–361.
Moore, Chris and Corkum, Valerie. 1994. "Social understanding at the end of the first year of life." *Developmental Review* 14:349–372.
Pedersen, Janni, Segerdahl, Pär and Fields, William M. 2009. "Why apes point: Pointing gestures in spontaneous conversation of language-competent *Pan/Homo* bonobos." In *Primatology: Theories, methods, and research,* Emil Potocki and Juliusz Krasinski (eds), 53–74. Hauppauge, NY: Nova.
Pika, Simone and Mitani, John. 2006. "Referential gestural communication in wild chimpanzees (*Pan troglodytes*)." *Current Biology* 16:R191–R192.
Povinelli, Daniel J., Bering, Jesse M. and Giambrone, Steve. 2003. "Chimpanzee 'pointing': Another error of the argument by analogy?" In *Pointing: Where language, culture, and cognition meet,* Sotaro Kita (ed.), 35–68. Hillsdale, NJ: Erlbaum.
Racine, Timothy P., Leavens, David A., Susswein, Noah and Wereha, Tyler J. 2008. "Pointing as intersubjectivity in human and nonhuman primates." In *Enacting intersubjectivity: A cognitive and social perspective to the study of interactions,* Francesca Morganti, Antonella Carassa and Giuseppe Riva (eds), 65–79. Amsterdam: IOS Press.
Rodriguez, Cintia. 2009. "The 'circumstances' of gestures: Proto-interrogatives and private gestures." *New Ideas in Psychology* 27:288–303.
Rumbaugh, Duane M. (ed.). 1977. *Language learning by a chimpanzee: The Lana project.* New York: Academic Press.
Savage-Rumbaugh, Sue E., MacDonald, Kelly, Sevcik, Rose A., Hopkins, William D. and Rubert, Elizabeth. 1986. "Spontaneous symbol acquisition and communicative use by pygmy chimpanzees (*Pan paniscus*)." *Journal of Experimental Psychology: General* 115:211–235.

Savage-Rumbaugh, Sue E., Rumbaugh, Duane M. and MacDonald, Kelly. 1985. "Language learning in two species of apes." *Neuroscience & Biobehavioral Reviews* 9:653–665.
Savage-Rumbaugh, Sue, Shanker, Stuart G. and Taylor, Talbot J. 1998. *Apes, language, and the human mind.* Oxford: Oxford University Press.
Savage-Rumbaugh, Sue, Wilkerson, Beverly J. and Bakeman, Roger. 1977. "Spontaneous gestural communication among conspecifics in the pygmy chimpanzee (*Pan paniscus*)." In *Progress in Ape Research*, Geoffrey Bourne (ed.), 97–116. New York: Academic Press.
Searle, John R. 1975. "A taxonomy of illocutionary acts." In *Language, mind, and knowledge, Minnesota Studies in the Philosophy of Science*, Vol. 7, Keith Günderson (ed.), 344–369. Minneapolis, MN: University of Minnesota Press.
Skinner, Burrhus F. 1957. *Verbal behavior.* Action, Massachusetts: Copley Publishing Group.
Tanner, Joanne E. and Byrne, Richard W. 1996. "Representation of action through iconic gesture in a captive lowland gorilla." *Current Anthropology* 37:162–173.
Tomasello, Michael, Carpenter, Malinda and Liszkowski, Ulf. 2007. "A new look at infant pointing." *Child Development* 78:705–722.
Tomasello, Michael, George, Barbara L., Kruger, Ann C., Farrar, Michael J. and Evans, Andrea. 1985. "The development of gestural communication in young chimpanzees." *Journal of Human Evolution* 14:175–186.
Veà, Joaquim J. and Sabater-Pi, Jordi. 1998. "Spontaneous pointing behaviour in the wild pygmy chimpanzee (*Pan paniscus*)." *Folia Primatologica* 69:289–290.
Wilkins, David. 2003. "Why pointing with the index finger is not a universal (in sociocultural and semiotic terms)." In *Pointing: Where language, culture, and cognition meet,* Sotaro Kita (ed.), 171–215. Hillsdale, NJ: Erlbaum.
Witmer, Lightner. 1909. "A monkey with a mind." *The Psychological Clinic* III:179–205.
Woodruff, Guy and Premack, David. 1979. "Intentional communication in the chimpanzee: The development of deception." *Cognition* 7:333–362.
Xitco, Mark J., Gory, John D. and Kuczaj, Stan A. 2001. "Spontaneous pointing by bottlenose dolphins (*Tursiops truncatus*)." *Animal Cognition* 4:115–123.
Xitco, Mark J., Gory, John D. and Kuczaj, Stan A. 2004. "Dolphin pointing is linked to the attentional behavior of a receiver." *Animal Cognition* 7:231–238.
Yerkes, Robert M. 1943. *Chimpanzees: A laboratory colony.* New Haven: Yale University Press.

Requesting behaviors within episodes of active sharing

A new look on chimpanzee signaling

Ray Wilkinson,[1] Ivan Leudar[1] & Simone Pika[2,3]
[1]University of Manchester, School of Psychological Sciences, Manchester, UK /
[2]Humboldt Research Group "Comparative Gestural Signalling",
Max Planck Institute for Ornithology, Seewiesen, Germany
[3]Corresponding author

Similarly to humans, one of human's closest living congener, chimpanzees (*Pan troglodytes*), have been observed to share food between unrelated individuals regardless of age and sex. The most remarkable cases of extended food sharing episodes including active sharing occur after chimpanzees have preyed upon mammals such as monkeys, duikers or bush pigs. Although a tremendous amount of research attention has focused on the function underlying the sophisticated sharing events, relatively little systematic research has investigated the behaviors and communicative signals surrounding and initiating these events. In the present paper, we thus provide the first systematic analysis of active meat sharing episodes by combining methods of comparative research with a micro-analytic approach in the form of conversation analysis. We describe how chimpanzees involved in sharing episodes achieve active sharing through a three-part process of (1) B requesting meat; (2) A giving meat; and (3) B taking meat. In addition, we describe the interactive process by which requests may not be acceded to and how subsequent requests may then be shaped and customized.

1. Introduction

Food sharing consists of one individual relinquishing control over a food item in favor of another and is ubiquitous among human cultures (Gurven 2004). Due to its apparent altruistic nature and its possible impact on the evolution of cooperative skills, food sharing has received considerable research attention by a variety of scientific disciplines (e.g. Heeschen, Schiefenhövel & Eibl-Eibesfeldt 1980; Hill, Kaplan & Hawkes 1993; Isaac 1978; Stevens 2004). Behavioral biologists for instance focus on discovering both the ultimate mechanisms and the proximate determinants that underlie and shape this and related costly prosocial behaviors (Caporael, Dawes, Orbell & Van de Kragt 1989; Winterhalder 1996).

Anthropologists investigate the pattern of food transfer among small-scale subsistence economies (e.g. Kaplan et al. 1985), and the impact of food sharing on the development of sophisticated cognitive skills (Parker & Gibson 1979).

Food sharing has also been reported in a variety of other animal taxa and species such as insects (Bolten, Robinson, Nation & Yu 1983; Vahed 1998), birds (Green & Krebs 1995; Heinrich 1988; Scheid, Schmidt & Noe 2008; Wright & Leonard 2002), cetaceans (Hoelzel 1991; Johnson 1982), social carnivores (Mech 1970; Schaller 1972) and vampire bats (Denault & McFarlane 1995; Wilkinson 1992). Concerning our closest living relatives, the non-human primates, (hereafter primates), food sharing is especially developed in omnivorous species (Feistner & Price 1990; Ferrari 1987; Perry & Rose 1994) and most exclusively restricted to offspring or other members of the same reproductive unit (Brown, Almond & van Bergen 2004; Feistner & McGrew 1989).

Interestingly however, similarly to humans (Stevens & Gilby 2004; Wilson 1975), chimpanzees (*Pan troglodytes*) have been observed to also regularly share food between unrelated individuals of various age and sex classes (Goodall 1986; Teleki 1973). The most remarkable cases of extended food sharing occur after chimpanzees have preyed upon mammals ranging in size from small rodents, over monkeys to juvenile bush pigs (for an overview see, Wrangham, Van Zinnicq & Riss 1990; see Figure 1).

Figure 1. Chimpanzee BS chewing on a bone of a previously hunted red colobus monkey. ©Langergraber

To understand the function underlying these interactions, a variety of hypotheses have been proposed (e.g. Gilby 2006; Gomes & Boesch 2009; Mitani & Watts 2001; Wrangham 1975). Surprisingly however, relatively little research attention has been paid to the behavior and communicative signals surrounding and initiating these sharing events. To date, early reports for instance mention behaviors and communicative gestures[1] by individuals who try to obtain meat from possessors, e.g. BEGGING WITH OUTSTRETCHED PALM,[2] COVERING THE POSSESSOR'S MOUTH WITH ONE OR BOTH HANDS, PEERING, PULLING ON THE CARCASS, REACHING OUT TO TOUCH THE MEAT OR THE CHIN AND LIPS OF THE MEAT OWNER, and accompanying vocalizations such as soft 'whimpers' or 'hoo' calls (Goodall 1986; Takahata, Hasegawa & Nishida 1984; Teleki 1973; Wrangham 1975). Feistner and McGrew (1989) describe the behavior of a meat possessor, who TURNS AWAY to avoid a female's attempt to obtain a piece of meat and RAISES his right ARM to fend her off.

To account for the huge variability of behaviors food sharing can employ, Boesch and Boesch (1989; p 551) distinguished between six levels of increased sharing intention on behalf of the food owner:

Theft	B *forcefully takes* part or all of A's food.
Recovery	B *takes* part of the food that A has dropped on the ground or placed there.
Passive sharing	B *takes* part of the food that A is holding. A makes no movement to facilitate nor to avoid B's action.
Active-passive sharing	B *takes* part of the food that A is holding. A makes a movement to facilitate B's action.
Active sharing	A *gives* part of its food to B, by either cutting off a piece or by holding a piece towards A.
Gift	A *gives* a part at least three times larger than its own remaining piece to B.

Furthermore, Gilby (2006) differentiated between four types of begging events: (1) *sit* (sitting and staring at the carcass), (2) *reach* (reaching towards but not touching the carcass or possessor), (3) *hold* (touching the carcass or possessor) and (4) *hand-mouth* (placing a hand directly over the possessor's mouth).

Overall, however, no detailed analyses have been carried out focusing on the communication strategies used to request meat and the communicative interplay between signaler and signal receiver. Recently however, we started a long-term project on gestural signaling in chimpanzees in the wild (Pika & Mitani 2006; Pika & Mitani 2008) with one special focus being meat sharing events and the underlying cognitive complexity that these involve (Pika & Mitani forthcoming).

---

1. These reports however do not properly define the term gesture or follow a specific gesture description (for definitions see, Pika 2008b, 2009)

2. Gestures and perlocutionary signals (see paragraph 2.6) described in this paper will be denoted in small capitals.

What is clear so far from available analyses of interaction is that particular signals obtain their meaning from their context and in particular their position in the flow of behavior of the participants. To understand sharing interactions, it is therefore necessary to describe both the form of behavioral signals and where these signals occur within the unfolding course of social action produced by the chimpanzees.

The present paper represents the first analysis from our data on sharing episodes. It is based on methods of comparative research (applying a combination of ethology (e.g. Van Hooff 1973; Van Lawick-Goodall 1968) and comparative Psychology (Call & Tomasello 2007) with a micro-analytic approach based predominantly on conversation analysis methods and findings (ten Have 2007).

Since conversation analysis (CA) is a new approach within the field of comparative research, we will provide some brief introductory comments. CA is a primarily qualitative form of micro-analysis of human action which emerged from various research strands within the social sciences, in particular ethnomethodology (Garfinkel 1967) and the work of Erving Goffman (e.g. Goffman 1964). To our knowledge CA has not been systematically applied to behavioral interactions between primates in previous published research.

The main feature of conversation analytic work in humans is its focus on social actions (such as requests, offers, questions, greetings etc.), including (i) the forms these take, (ii) their sequential positioning within the larger ongoing social activity, and (iii) how they are responded to (or not) by recipients (Sidnell 2010). In human interaction, these actions and their uptake are regularly produced within turns-at-talk, which are organized such that one speaker talks at a time, holding the floor as the other speaker or speakers remain largely silent (Sacks, Schegloff & Jefferson 1974). The recognisability of an initiating action, such as a request, is typically achieved through the use of certain linguistic forms (e.g. Curl & Drew 2008). A common way in which actions and the responses to them are organized is through so called 'adjacency pairs', comprising a sequence of two turns produced by different speakers (Schegloff 2007). Adjacency pairs consist of an initiating action such as a request, and a responding action, such as an acceptance or rejection of that request. Adjacency pairs are normative; i.e. on the production of the initiating action a responsive action is due by the addressed speaker. The lack of such a response means it can be seen to be 'officially absent'. This absence may engender further activity in relation to that absence, such as the speaker of the initiating action re-doing that action, or by the 'negligent' speaker providing an account for why the expected action was not produced (Heritage 1984).

In combining the comparative approach to gestural signaling (e.g. Pika 2008a; Call & Tomasello 2007; based also on some terms developed by Bates and colleagues 1975, 1979) with a micro-analytic approach in the form of conversation analysis, this chapter has the following two aims:

1. To provide detailed insight into active sharing events of chimpanzees (following the definition of Boesch & Boesch 1989) with a special focus on request behaviors (i.e. here, displays of interest in gaining meat) and the interactions between chimpanzees which lead from the request to the subsequent transfer of meat;
2. To offer a new approach to the understanding of gestural communication in chimpanzees, by focusing on request gestures within episodes of active sharing of meat.

We therefore intend to provide the first close (descriptive) analysis of both (a) the form of request gestures in chimpanzees, alongside (b) the sequential location in which they occur within the activity and how they are responded to.

2. Methods

For the current analysis, we used data on meat sharing events between adult chimpanzee males collected by SP from June-August 2008 at the Ngogo community, Kibale National Park, Uganda. Detailed descriptions of the Ngogo study area can be found in Struhsaker (1997) and Butynski (1990).

2.1 The Ngogo chimpanzee community

The Ngogo community is the largest community that has been described in the wild. It contained approximately 150 individuals, including 25 adult males, 14 adolescent males, and 44 adult females, during the period considered here. Ngogo chimpanzees mostly hunt red colobus monkeys (*Procolobus rufomitratus*), and are more successful at hunting and make more kills per hunt than chimpanzees elsewhere (Mitani & Watts 1999). We will refer in the present paper to distinct individuals by using two-letter abbreviations of their names, e.g. BT for Bartok.

The Ngogo community is an ideal place to study behavioral interactions and gestures, since continuous observations of the chimpanzees have been maintained since 1995, and most animals are well habituated to the presence of human observers.

2.2 Behavioral observations

SP followed adult male chimpanzees of the Ngogo community on 48 days between 07.00 a.m. and 05.00 p.m. She was able to record meat sharing events following six different hunts on six different days using a digital camera (Panasonic NV-GS 250; sampling rule: behavior sampling; recording rule: continuous recording) and a digital recorder (Audioline, VR 500). The three examples analysed here were chosen from the larger set of active sharing episodes in these data. We chose examples which (a) were representative of such behaviors in the wider data set; (b) involved the (eventual) transfer of meat, providing some evidence that the gestures we were analyzing as requests for meat were also treated that way by the meat-holding chimpanzee; and (c) where the behaviors of all the chimpanzees within the episode of meat sharing were relatively visible on the video recording, allowing us to transcribe and analyse them.

2.3 Application of conversation analysis

In keeping with conversation analytic methodology, our interest and focus in this chapter is on how chimpanzees achieve active meat sharing through their behavior in interaction with each other on a moment-by-moment basis. In particular we focus on (a) the positioning of the 'requesting' chimpanzee in relation to the meat-holding chimpanzee such that a request will be visible; (b) the types of gestures used; (c) how requests are either acceded to (by transferring meat) or not acceded to (possibly leading to a further request); (d) how the smooth transfer of meat is achieved; and (e) in particular how certain actions by the meat-holding chimpanzee may be recognizable as displaying that meat may be about to be transferred, allowing other chimpanzees to get into a position to receive that piece.[3]

Conversation analysts typically provide quite detailed transcriptions of talk and other forms of behavior in interaction, such as gesture and its relation to the surrounding talk (Goodwin 2007; Schegloff 1984), to both facilitate analysis and present that analysis to readers. In the extracts below, we provide a description of chimpanzee behaviors, which aims to capture the moment-by-moment unfolding of those behaviors both by individual chimpanzees and between two or more chimpanzees. To allow a more reader-friendly précis of the relevant points, we also provide a summary of request- and sharing-relevant behaviors in a second column.

Our project here may be thought of as 'CA-assisted comparative research' (cf. Heeschen & Schegloff 1999: 378) on 'CA-assisted aphasiology'), where analytical questions and prior findings from comparative research are combined with a CA approach to the description and analysis of courses of action between chimpanzees; –

3. We are aware of the dangers of anthropomorphizing and mis-representing these behaviors through the use of terms such as 'request' and 'gesture', and retain these terms due to their prevalence in the primate literature. It should be noted that the range of behaviors we refer to with these terms is generally wider than that used when these terms describe human behaviors. For example, 'request' here covers behaviors such as gazing intently at the meat (which may be better described as 'begging'). Also, our use of 'gesture' here can refer to chimpanzee body-behavior generally and is not limited to communicative manual gesture (for the definition of gesture see, Pika 2008b, Scott & Pika this volume).

in particular gestural requests and responses to those requests. We now turn to an outline of the types of communicative behaviors focused on in the analysis, derived from previous comparative research as well as inductively from the interactional data of chimpanzees we have been exploring and which we report on here.

2.4 Gesture definition

Following the definition of Pika (2008b), a gesture is defined as movements of the limbs or head and body postures that are directed toward a recipient and are mechanically ineffective (which means, they are not designed to act as direct physical agents). An important condition is that request-signs elicit an appropriate *uptake* – a voluntary response from the addressee (cf. Austin 1960). This condition ensures that the request-sense is not arbitrarily imposed on the behavior by the analysts. The following behavioral criteria were used to infer their communicative intent: (1) gazing at the recipient of a given signal, and/or (2) waiting for a possible response after the signal had been produced.

In addition, in our descriptions of Extracts, we are using the following gesture types, based on our ongoing work on communicative signaling during meat sharing events (Pika & Mitani forthcoming):

2.5 Gestures

2.5.1 *Gestures in close proximity (0–15 cm)*

DIRECTED DROP: The possessor places a piece of meat in the hand of a distinct recipient.

HAND CUPPED: Hand is hollowed and placed underneath the hand of the possessor or the meat.

HOLD ONTO: The animal grasps onto the meat quite loosely, no pulling or pushing involved.

HOVER: The hand is circling around or held in a distance of circa 5–15 cm to the meat, palm directed towards it, without any direct contact.

OFFER: The animal holds out his hand and proffers a piece of meat to another individual.

PALM-UP: The animal stretches out the hand towards the meat possessor/meat, with the palm showing.

TOUCH: Gentle and short (< 5 sec) contact with flat hands, body part, or feet.

TOUCH GENITALIA: The animal is making gentle and short (< 5 sec) contact with the genitalia (penis, swelling, scrotum) of the recipient.

2.5.2 *Distant gestures (>15 cm)*

LEANING BACKWARDS: The animal is sitting and bends his upper body backwards in an angle of $>90^0$ (normal vertical sitting posture) but $\leq 150^0$.

HAND BENDED: The hand and arm are held in a vertical plane, with the back of the hand directed to the meat, wrist and elbow are bended, the fingers are positioned horizontally in a loose, relaxed manner.

PALM-UP: Definition see above.

PRESENT: The animal is showing a distinct body part (e.g. genitals) to the recipient.

REACH OUT: The arm, hand and fingers are stretched out toward the meat/possessor.

ROCKING: The animal moves his body back and forth in a rhythmic fashion.

2.6 Perlocutionary signals

In addition to gestures, which can be classified as illocutionary acts, because the signaler directs its behavior intentionally toward a recipient (Bates et al. 1975), we also coded perlocutionary acts. In perlocutionary acts, the communication occurs only because the receiver is adept at interpreting the behavior of the signaler (Bates et al. 1975).

DIVIDE: The possessor breaks off a piece from the meat or uses pulling, gnawing and tearing movements to divide the meat in two pieces.

POSITION IN INNER TRIANGLE: The beggar creates a virtual triangle with the possessor and the meat, with the meat in touching distance to the beggar and not shielded by the arms of the possessor.

TURN AWAY: The possessor averts from the beggar with the meat in an angle of circa 30–90^0, and restricts his access to the meat.

TURN BACK: The possessor averts from the beggar in an angle of circa 91–180^0, and restricts his reaching and visual access.

3. Results

We present below the analysis of three episodes of active sharing of meat. In each, the transfer of meat is preceded by a chimpanzee requesting (displaying an interest in) meat from a meat-holding chimpanzee. Our interest here is in exploring the moment-by-moment behaviors and responses to those behaviors involved in these episodes to explore the interactional processes through which active sharing of meat is carried out.[4]

4. While in the present chapter we do not carry out quantitative analyses, we expect that the phenomena we investigate qualitatively in this chapter will form the basis for such analyses in future work.

Extract 1a displays some of the behaviors seen recurrently in the dataset through which a chimpanzee may position himself in relation to the meat-holding chimpanzee to request meat.[5] It also displays how that request may not be acceded to and what the 'requesting' chimpanzee may then do to produce a further request.[6]

3.1 Extract 1a (M2U00894_00:00–01:34)

(Chimpanzees MT [Morten] and MW [Mwea; missing the left hand] are in the branches of a tree. There is another chimpanzee, BS [Basie], who enters the scene as the episode progresses and sits on a higher branch, but appears to take no active part in this episode. MT is holding the torso of a recently hunted sub-adult red colobus monkey (*Procolobus rufomitratus*) in his hands and is eating it).

Time (in sec)	Line	Description of behaviors	Summary of request- and sharing-relevant behaviors
00:00–00:16	1	MT and MW are broadly facing each other with MT holding the meat in front of him and eating from it. MW is displaying interest in the meat by gazing at MT/the meat. It is not possible to tell from the camera angle whether he is also displaying interest in the meat by using gestures.	MW gazing at MT/the meat.
00:16–00:18	2	MT turns to his right, TURNING AWAY from MW with his body now between MW and the meat.	MT TURNS AWAY.
00:18–00:21	3	MW POSITIONS himself parallel to the right of MT IN the INNER TRIANGLE and is in sight of MT, and within touching distance to the meat.	MW POSITIONS IN INNER TRIANGLE and faces MT/the meat.
00:21–00:23	4	As MW starts to REACH OUT his arm towards the meat, MT immediately again TURNS AWAY further to his right, again taking the meat out of MW's reach.	MW REACHES OUT arm towards meat; MT TURNS AWAY.

(*Continued*)

5. The transcripts here consist of four columns. Column 1 displays the time on the video clip from which the extract is taken. Column 2 is a line number which allows a means of reference to a particular part of the transcript. Column 3 is a relatively detailed descriptive record of the behaviors in the episode. Column 4 provides a summary of request- and sharing-relevant behaviors.

6. Due to their length, Extract 1 and Extract 2 are presented and discussed in two parts (Extract 1a, 1b, 2a, 2b).

Time (in sec)	Line	Description of behaviors	Summary of request- and sharing-relevant behaviors
00:23–00:25	5	MW REACHES OUT his hand towards the meat (now probably out of his reach) and MT then takes a step away from MW along the branch and TURNS AWAY.	MW REACHES OUT HAND towards meat; MT TURNS AWAY.
00:25–00:33	6	After staying in the same position for a couple of seconds, MW then moves along a branch past MT and ends up slightly above MT, POSITIONING himself IN the INNER TRIANGLE, again within touching distance of the meat and potentially in MT's line of vision.	MW POSITIONS himself IN INNER TRIANGLE.
00:34–00:55	7	For around 20 seconds, MW HOVERS his hand around the meat while MT is holding it and bites from it.	MW HOVERS his hand around the meat.
00:56–01:00	8	As MW's REACHES OUT once again, MT pulls the meat away slightly to his left, TURNS AWAY from MW and continues to bite it. MW ROCKS back and forth, vocalizes, showing a 'bared-teeth display' (Goodall 1986), LEANING BACKWARDS and PRESENTING his genitals/belly, while MT continues to bite the meat.	MT TURNS AWAY/pulls meat away from MW.
01:00–01:01	9	MT TOUCHES MW's GENITALIA with his right hand, and MW again REACHES OUT towards the full piece of meat in MT's hand.	
01:01–01:05	10	As MW REACHES OUT towards the meat, MT again TURNS further to his left, AWAY from MW.	MW REACHES OUT towards meat; MT TURNS AWAY.
01:05–01:07	11	MW steps to the left down the branch following MT's direction and REACHES OUT back and forth towards the meat, HOVERS over it but MT again TURNS further round to the left AWAY from MW.	MW REACHES OUT towards meat, HOVERS over; MT TURNS AWAY.
01:08–01:13	12	MW again moves along the branch past MW and POSITIONS himself again IN the INNER TRIANGLE directly in front of the meat, which is now between the two individuals.	MW POSITIONS himself IN the INNER TRIANGLE of MT/the meat.
01:13–01:22	13	MT MOVES slightly back along the branch in the other direction AWAY from MW but stops with the meat held in front of him and biting from it. MW requests some meat by REACHING OUT and stretches out his hand PALM-UP under it.	MT TURNS AWAY from MW. MW uses PALM-UP gesture under meat.

(Continued)

Time (in sec)	Line	Description of behaviors	Summary of request- and sharing-relevant behaviors
01:22–01: 34	14	MT TURNS AWAY to his right, still eating the meat; MW withdraws his hand, trying to find his balance on the branch, scratches himself under his left arm, then resumes gazing at MT eating.	MT TURNS AWAY. MW gazing at MT eating meat.

In Extract 1a chimpanzee MW uses his hands in ways which do not involve attempts to grab or take the food in an instrumental way, but rather which communicate to chimpanzee MT an interest in being given some of the meat. One form this takes, is to place his hands in close proximity to the meat but not to touch it. In Extract 1a, for instance, MW on several occasions REACHES OUT towards the meat (lines 4, 5, 8, 10 and 11). He also HOVERS his hand around the meat (line 7) and CUPS his HAND under it (line 13). An example of requesting meat by REACHING OUT towards it can be seen in Figure 2.

In each case, these behaviors are designed to display an interest in the meat. Another way in which interest is displayed is through gazing at the meat owner and the meat (line 1). A regular response from MT following a request/display of interest from MW is to TURN his body and/or orientate the meat in a direction AWAY from MW's reach. In these cases MT therefore does not respond by providing any meat and, as such, does not accede to MW's request/display of interest. In Extract 1a, this occurs in lines 2, 4, 5, 8, 10, 11, 13 and 14.

Figure 2. MW (on left of picture) requests meat by REACHING OUT his hand towards it as it is held by MT (on right of picture) (Extract 1a, line 8).

So far we have seen that when MW requests/displays interest in the meat, MT TURNS AWAY, withdrawing the meat from MW's reach. What, if anything, does MW do in response to this? Regularly in this extract, he moves and repositions himself in a place where he is likely to be within MT's line of vision (POSITION IN INNER TRIANGLE). He does this three times (lines 3, 6 and 12). This position within the line of vision can be seen to be important where the communication is in a viso-spatial form, such as gesture and other forms of body behavior, and has to be seen by the recipient in order to be effective.

A recurrent set of behaviors observed in Extract 1a, therefore, is that of chimpanzee MW requesting, chimpanzee MT not acceding to that request and instead TURNING AWAY, and MW then repositioning himself in front of MT such that another request might be launched.

Extract 1b displays the transfer of meat and the behaviors by the meat-holding chimpanzee (MT) preceding it, which appear to be treated by MW as indicating that possible transfer is imminent.

3.2 Extract 1b (continuation of 1a) (M2U00894_01:34–01:50)

Time (in sec)	Line	Description of behaviors	Summary of request- and sharing-relevant behaviors
01:34–01:42	15	MT starts to pull the carcass apart through biting and pulling with greater force than immediately previously. He holds it in both hands and gnaws at it forcibly. MW's body demeanour changes; he stands up from his poised position and moves back and forth slightly as if unsure whether to move forwards or not, or waiting for the right time to move forward.	
01:43–01:46	16	As MT continues to bite at the meat, MW moves forward quite slowly along the branch. His hand brushes/taps MT's shoulder as he moves past. He POSITIONS himself IN the INNER TRIANGLE in MT's line of vision, in a position to accept meat if it is offered.	
01:46–01:48	17	MT DIVIDES the meat and holds out a piece with his left hand into the space below MW. MW quickly extends his right hand down and takes the meat which MT is OFFERING.	MT OFFERS meat and MW, in position, takes it.
01:48–01:50	18	Immediately after transfer, MW moves away up the branch, and MT TURNS AWAY in the other direction with the remaining meat.	

When meat is eventually transferred, it is achieved smoothly i.e. as MT DIVIDES the meat and OFFERS a piece, MW is already in position to stretch out his hand and to receive that piece (line 17). That smooth transfer can be seen to be the result of some perlocutionary behavior by MT, which is recognizable to MW as foreshadowing the possible transfer of meat. As MT continues to bite the carcass, he also starts to pull it apart with greater force than previously. As he holds it in both hands, he gnaws at it, loosening some of the meat from the carcass. MW's whole body comportment now changes (line 15). He appears to be treating MT's biting and pulling apart of the carcass as not simply another stage in MT's activity of eating the meat himself, but rather as a possible precursor to that meat being transferred. He stands up, tries to find his balance on the branch, and waits in the same location, observing MT. He then moves quite slowly down the branch past MT. His body is now in a position ready for transfer (i.e. to the side of/slightly in front of MT) (line 16). As MT then appears to DIVIDE the meat to OFFER a piece, MW is in a position to REACH OUT and take it (line 17).

Similarly in Extracts 2a and 2b, perlocutionary acts and gestural displays of interest in receiving meat lead to a transfer of meat. Here there are two chimpanzees (Harelip [HL] and Mwea [MW]) requesting meat.

3.3 Extract 2a (M2U00821_04:19–05.00)

(Chimpanzee BT [Bartok], with the torso of a recently hunted red duiker *(Cephalophus natalensis)* in his hands, is sitting on the branch of a tree. To his right, slightly lower down the same branch, is chimpanzee HL [Harelip]. Chimpanzee MW [Mwea] is on a slightly higher branch facing chimpanzee BT with the meat in between them. For around 18 seconds in this episode, both HL and MW have been displaying interest in the meat by either HOVERING a hand near it or using the hand to lightly TOUCH it.

Time (in sec)	Line	Description of behaviors	Summary of request- and sharing-relevant behaviors
04:19–04:22	1	HL moves from behind BT, where he sat, down the branch and POSITIONS himself IN the INNER TRIANGLE in front of BT, with the meat now in between them, his right HAND BENDED.	HL POSITIONS himself IN INNER TRIANGLE towards BT/the meat.
04:23–04:44	2	HL HOVERS over/around the meat, perhaps sometimes lightly TOUCHING it. After about ten seconds (at 04:33) MW's hand descends and starts to HOVER over and, on occasion, lightly to TOUCH the meat.	HL HOVERS his hand over the meat/lightly TOUCHES it.

(Continued)

Time (in sec)	Line	Description of behaviors	Summary of request- and sharing-relevant behaviors
04:45–04:50	3	HL withdraws his right hand from the meat and puts it behind him (it appears he may perhaps be temporarily clasping onto a branch out of camera shot to steady himself). He starts instead to REACH OUT with his left hand towards the meat.	HL REACHES OUT towards the meat.
04:50–04:51	4	As HL HOVERS over and towards the meat, BT is visibly starting to shift into a posture, which will allow him to bite a piece off the meat; he shifts position slightly TURNS AWAY to his left (i.e. away from HL) and starts to bring his right hand up onto an already half-torn off part of the meat, which is also now being grabbed and pulled quite firmly by MW. The meat is in front of BT's mouth in a position from which a piece can be gnawed off by BT.	BT TURNS slightly AWAY from HL. MW and HL grabbing the meat.
04:51–04:54	5	As BT is shifting into the position described above, HL quickly changes his physical orientation. He retracts his left hand, extends his right hand and takes a step forward up the branch to POSITION himself again IN the INNER TRAINGLE. He then HOVERS his right hand over the meat, sometimes also TOUCHING the meat alongside, and sometimes on top of, TOUCHING BT's hand.	HL POSITIONS himself IN the INNER TRIANGLE, HOVERS hand over meat and TOUCHES the meat and TOUCHES BT's hands on it.
04:54–04:59	6	A hand of each chimpanzee is now HOLDING ONTO the meat while BT gnaws at it.	
04:59–05:00	7	While DIVIDING the meat, BT is biting it and pulling it apart with both hands. This appears to have the effect of loosening a chunk of meat from the main body of the animal. As part of this process, BT pushes the hands of both HL and MW off of the meat.	BT DIVIDES the meat and pushes HL and MW's hands off of the meat.

In Extract 2a, a request for meat again takes the form of a chimpanzee HOVERING his hand towards it (line 2). In this episode, the request/display of interest also takes another form, that of TOUCHING the meat or the hand of the possessor, which is touching the meat. In lines 4 to 6, MW is displaying his desire to have some meat by TOUCHING it as BT prepares, and then proceeds, to gnaw it. HL, in competition with MW to receive the meat, responds to this by moving forwards and TOUCHING BT's hand as BT's hand holds the meat. As with the behaviors seen in Extract 1a, these behaviors do not appear to function instrumentally to try to take the meat but rather act as true gestures to communicate interest in the meat as precursors to being offered some.

The transfer of meat and the behaviors which precede it are seen in Extract 2b.

3.4 Extract 2b (continuation of 2a) (M2U00821 05:00–05:03)

Time (in sec)	Line	Description of behaviors	Summary of request- and sharing-relevant behaviors
05:00–05:01	8	BT forcefully pushes the meat away from himself again, leaving a piece in his mouth.	
05:01–05:02	9	At the same time as BT reaches towards his own mouth, both HL's hand and MW's hand start to HOVER in the direction of BT's moving hand. As BT takes the meat out of his mouth, both HL's hand and MW's hand are close to BT's hand, in a position to receive the meat. HL CUPS his HAND, while MW's hand is HOVERING above. BT uses a DIRECTED DROP to transfer meat to HL.	HL's hand and MW's hand both HOVER around BT's hand containing meat, ready for transfer. BT uses a DIRECTED DROP and gives meat into HL's CUPPED HAND.
05:02	10	Immediately after transfer, the close huddle breaks up. HL steps back down the branch while putting the meat in his mouth and grabbing some leaves from the branch to eat with the meat, BT reaches up to a branch and pulls down some leaves to eat with his meat. MW retracts his hand.	

Here, as was also seen in Extract 1b, chimpanzee BT's DIVIDING the meat (involving biting and pulling apart of the meat) is an observable action, which is treated by other chimpanzees as indicating that a transfer of meat may possibly be imminent. As BT has pulled off a piece of meat with his mouth and is reaching towards his mouth to retrieve it, both HL's hand and MW's hands start to move towards BT's mouth/moving hand (Extract 2b, line 9). At this point, therefore, both can be seen to be treating BT's action as preliminary to the retrieval of meat from his mouth and the subsequent transfer to one of them. As BT takes the meat from his mouth, both HL's hand and MW's hand are extended, located near BT's hand in a position to receive the meat. BT reaches out and carries out a DIRECTED DROP in HL's CUPPED HAND (line 10).

Finally, in Extract 3 (involving three chimpanzees), gestures requesting meat are again followed by a transfer of meat. Here, the request (by chimpanzee MW) involves again a large submissive gesture (see also Extract 1a, line 8).

3.5 Extract 3 (M2U00823 03:22–03:38)

(As the recording clip starts, three chimpanzees are sitting together in a loose triangle formation. Chimpanzee BT [Bartok], on the left, has the torso of a red duiker *(Cephalophus natalensis)* in his left hand and is eating. Chimpanzee MW [Mwea], in the centre, has his head lowered and is gazing at BT's meat. Chimpanzee DP [Dolphi], on the right, is facing into the inner triangle but appears to be taking no active part. He is eating something.)

03:22–03:25	1	MW approaches with head lowered and left forearm shielding his face. He grabs the meat using his right hand while glancing up at BT.	MW grabs the meat.
03:25–03:30	2	As BT moves the meat, MW retains his hand on it quite passively i.e. not apparently pulling at the meat, but being pulled by it as BT moves it.	
03:30–03:31	3	BT brings his right hand across and transfers the meat to his right hand, lifting the meat nearer to, and in front of, himself. MW's hand is removed from the meat, either passively – by BT moving the meat – or actively – by BT's hand removing MW's (it is not possible to tell which of these two possibilities from the camera angle).	BT moves meat away from MW.
03:31–03:32	4	MW vocalizes and shows briefly a 'bared-teeth display' while LEANING BACKWARDS and cowering behind his right shoulder/arm; gazing above shoulder at BT.	
03:33–03:34	5	BT raises left hand to his mouth, takes out food from there and OFFERS it to MW.	BT OFFERS meat to MW.
03:34–03:38	6	MW gazes down at food/BT's hand; takes meat; puts it in mouth and almost immediately puts right arm over his face, while gazing at BT.	MW takes offered meat.

In Extract 3, MW first requests meat by approaching BT (the meat possessor) with his head lowered and his forearm partly covering his face and putting a hand on the meat (line 1). He retains a hold of the meat for several seconds as it is manipulated by BT (line 2). MW's hand is then removed, either by the movement of the meat as BT transfers it to his right hand, or by BT actively removing it (the details are not visible on the video). At this point (line 4), MW engages in a large submissive gesture (Pika & Mitani forthcoming): he LEANS BACKWARDS in a type of cowering posture and vocalizes, briefly showing a 'bared-teeth display' (Goodall 1986), while gazing over his shoulder at BT (Figure 3). BT then takes a piece of meat from his mouth and OFFERS it to MW for MW to take (line 5), which he does (line 6).

Figure 3. Chimpanzee MW (center) requesting meat LEANING BACKWARDS, and gazing over his shoulder at BT while vocalizing and baring his teeth. At this point, chimpanzee BT (left) is starting to transfer the meat using his left hand (Extract 3, lines 4 and 5).

4. Conclusions

The main aim of the present paper was twofold: First, we wanted to provide a better understanding of the communicative signals (perlocutionary and illocutionary) involved in active sharing events of wild chimpanzees; and second we aimed to introduce a new research tool by adopting the comparative approach to gestural communication (Pika 2008a) in combination with conversation analysis (ten Have 2007). What has been discovered in this analysis?

Theft of food, as Boesch and Boesch (1989) define it, involves the forceful taking of another's food and no communication is logically necessary. Likewise, in *recovery,* one chimpanzee simply takes food that another has left on the ground. In the episodes of *active sharing* analysed here, on the other hand, we have examined how the transfer of meat comes about through a communicative process involving two or more chimpanzees. Boesch and Boesch (1989) define active sharing as 'A gives part of its food to B either cutting off a piece or by holding a piece towards A.' In the extracts examined here, we have examined not just the transfer, or giving itself but rather the larger process including, in these cases, one or more requests by a "requesting" chimpanzee which precede the giving by the meat-possessing chimpanzee, as well as perlocutionary acts.

The sharing process can be an extended one (see, for example, Extract 1) where a request by a chimpanzee is resisted by the meat possessor and a further request then ensues. A common way in which the meat possessor resisted the requests was by partly TURNING AWAY from the requester and thus at the same time moving the meat further away (Extract 1). When this happened, for a further request to be possible, the requester could reposition himself so that once again the meat was located between him and the meat possessor and he was in or near the meat possessor's line of vision (POSITIONING IN INNER TRIANGLE) (Extract 1). As such there may be one or more rounds of requests by chimpanzee B, non-acceded by chimpanzee A, and one or more further requests (sometimes involving repositioning) by chimpanzee B.

The requesting behaviors in these extracts take forms such as:

- placing the hand near to the meat but not touching it. This can include communicative gestures such as HOVERING the hand over or near the meat or CUPPING the HAND under the meat;
- TOUCHING the meat or TOUCHING the hand of the possessor, which is already on the meat;
- gazing at the meat;
- submissive gestures involving LEANING BACKWARDS, PRESENTING genitalia, vocalizing, and briefly showing a 'bared-teeth display' while gazing over the shoulder (Extract 3).

In the extracts analysed here, these gestures are used in ways, which make these behaviors recognizable to the meat-holding chimpanzee *as* requests/displays of interest in gaining some of the meat. These requests and the responses to them are, of course, not organized through turns at talk or through the types of sequences of action, such as adjacency pairs, used in human interaction. How these chimpanzee actions are organized interactively is an area for further research.[7]

The active sharing evident in these extracts involves at least three stages:

1. a request/display of interest by B in A's meat;
2. the passing of meat by A towards B, for example by OFFERING it to be taken (for example, Extract 1); and
3. the taking of the meat by B.

As has been seen here, an episode of requesting may be prolonged since one or more requests may not be immediately acceded to.

The transfer of meat itself can also be an interactive process. The behaviors engaged in by A (e.g. DIVIDING the meat), enable the chimpanzees, who have requested meat, to see that transfer may be imminent. They can then use this information to get into a position ready for the transfer of a piece of meat

7. Whether there is anything at all comparable in chimpanzee interaction to the normative force of the initiating actions in adjacency pairs is also an area for further work.

(see Extracts 1 and 2). It is the structurally projectable features of certain behaviors – such as the biting off of a piece of meat in the manner seen here – which can allow observing chimpanzees to 'predict' what might happen next and to act accordingly. This type of observation thus emphasizes that studying the nature of chimpanzee routine activities represents a useful tool to understand the underlying cognitive complexity of communicative interactions in greater detail. In addition, it also provides a predictable structure within which these cognitive abilities are deployed and presumably learned.

In summary, in this paper we have analysed active sharing episodes of chimpanzees with a particular focus on the gestures that are used to request meat. By combing two methodologies, we have been able to describe how the chimpanzees involved in these extracts achieve active sharing through a three-part process of (1) B requesting meat; (2) A giving meat; and (3) B taking meat. We have also described the interactive process by which requests may not be acceded to and how a further request may then be produced.

We have examined behaviors such as requests using gestures in terms of their production within particular contexts and activities (such as the, often prolonged, attempts to get some meat). We have also investigated them in relation to how they are responded to by other chimpanzees. For example, when one chimpanzee responds to another's request to share meat by TURNING AWAY and not sharing the meat, the first chimpanzee may pursue that course of action by producing a further request. This method of examining primate behavior therefore allows us to analyse how such behaviors emerge within particular contexts (for example, at a point when a request has not been responded to with a transfer of meat) and how at the same time that behavior may itself constitute a context, which others will subsequently act within (cf. Heritage 1989). Examining primate behaviors in terms of their design for certain contexts and certain recipients, as well as how those recipients respond to them, may allow us to better understand these behaviors, their significance for those for whom they are designed, and the underlying cognitive complexity.

Acknowledgments

We are grateful to J. C. Mitani and D. P. Watts, who encouraged SP to study gestural behavior of chimpanzees in the Ngogo community and shared their tremendous knowledge on chimpanzee behavior and field skills. We thank the Uganda Wildlife Authority (UWA), the Uganda National Council for Science and Technology (UNCST) and the Makerere University for permission to work at the Makerere University Biological Field Station. We are grateful to J. Lwanga, A. Tumusiime, L. Ndagizi, G. Mbabazi, and A. Magoba for invaluable assistance and support in the field and at the station. Thanks also to A. Giessl and J. Rae for helpful discussions and comments on an earlier manuscript.

References

Austin, John L. 1962. *How to do things with words.* New York: Oxford University Press.

Boesch, Christoph and Boesch, Hedwige. 1989. "Hunting behavior of wild chimpanzees in the Taï National Park." *American Journal of Physical Anthropology* 78: 547–573.

Bolten, Anita B., Robinson, F. A., Nation, J. L. and Yu, S. J. 1983. "Food sharing between honeybee (*Apis mellifera*) colonies in flight cages." *Journal of Apicultural Research* 22: 90–100.

Brown, Gilian R., Almond, Rosamunde E. A. and van Bergen, Yfke. 2004. "Begging, stealing and offering: Food transfer in nonhuman primates." *Advances in the Study of Behaviour* 34: 265–295.

Butynski, Tom M. 1990. "Comparative ecology of blue monkeys (*Cercopithecus mitis*) in high- and low-density subpopulations." *Ecological Monographs* 60: 1–26.

Call, Josep and Tomasello, Michael. 2007. *The gestural communication of monkeys and apes.* Mahwah, New York: Lawrence Erlbaum Associates.

Camerer, Colin and Thaler, Richard. 1995. "Anomalies: Ultimatums, dictators, and manners." *Journal of Economic Perspectives* 9: 209–219.

Caporael, Linnda R., Dawes, Robyn M., Orbell, John M. and Van de Kragt, Alphons J. C. 1989. "Selfishness examined: Cooperation in the absence of egoistic incentives." *Behavioral and Brain Sciences* 12: 683–739.

Curl, Traci and Drew, Paul. 2008. "Contingency and action: A comparison of two forms of requesting." *Research on Language and Social Interaction* 41: 129–153.

Denault, Lisa K. and McFarlane, Donald A. 1995. "Reciprocal altruism between male vampire bats, *Desmodus rotundus*." *Animal Behaviour* 49: 855–856.

Feistner, Anna T. C. and McGrew, William C. 1989. "Food-sharing in primates: A critical review." In *Perspectives in Primate Biology*, Praveen K. Seth and Swadesh Seth (eds), Vol. 3, 21–36. New Delhi: Today & Tomorrow.

Feistner, Anna T. C. and Price, Eluned C. 1990. "Food-sharing in cotton-top tamarins (*Saguinus oedipus*)." *Folia Primatologica* 54: 34–45.

Ferrari, Stephen F. 1987. "Food transfer in a wild marmoset group." *Folia Primatologica* 48: 203–206.

Garfinkel, Harold. 1967. *Studies in Ethnomethodology.* Englewood Cliffs, New York: Prentice Hall.

Gilby, Ian C. 2006. "Meat sharing among the Gombe chimpanzees: Harassment and reciprocal exchange." *Animal Behaviour* 71: 953–963.

Goffman, Erving. 1964. "The neglected situation." *American Anthropologist* 66: 133–136.

Gomes, Cristina M. and Boesch, Christophe. 2009. "Wild chimpanzees exchange meat for sex on a long-term basis." *Plos One* 4: e5116.

Goodall, Jane. 1986. *The chimpanzees of Gombe, Patterns of Behaviour.* Cambridge, England: The Belknap Press of Harvard University Press.

Goodwin, Charles. 2007. "Participation, stance and affect in the organization of activities." *Discourse and Society* 18: 53–73.

Green, David J. and Krebs, Elizabeth A. 1995. "Courtship feeding in ospreys *Pandion Haliaetus* – a criterion for mate assessment." *Ibis* 137: 35–43.

Gurven, Michael. 2004. "To give and to give not: The behavioral ecology of human food transfers." *Behavioral and Brain Sciences* 27: 543–559.

Heeschen, Claus and Schegloff, Emanuel A. 1999. "Agrammatism, adaptation theory, conversation analysis: On the role of so-called telegraphic style in talk-in-interaction." *Aphasiology* 13: 365–406.

Heeschen, Volker, Schiefenhövel, Wulf and Eibl-Eibesfeldt, Irenäus. 1980. "Requesting, giving, and taking: The relationship between verbal and nonverbal behavior in the speech community of the Eipo, Irian Jaya (West New Guinea)." In *The Relationship of Verbal and Nonverbal Communication*, Mary R. Key (ed.). The Hague: Mouton de Gruyter.

Heinrich, Bernd. 1988. "Food sharing in the raven, *Corvus corax*." In *The Ecology of Social Behavior*, C. N. Slobodchikoff (ed.), 285–311. San Diego: Academic Press.

Heritage, John. 1984. *Garfinkel and Ethnomethodology*. Cambridge: Polity Press.

Heritage, John. 1989. "Current developments in Conversation Analysis." In *Conversation: An Interdisciplinary Perspective*, Derek Roger and Peter Bull (eds). Clevedon: Multilingual Matters.

Hill, Kim, Kaplan, Hillard and Hawkes, Kristen. 1993. "On why male foragers hunt and share food." *Current Anthropology* 34: 701–710.

Hoelzel, A. Rus. 1991. "Killer whale predation on marine mammals at Punta Norte, Argentina: Food sharing, provisioning and foraging strategy." *Behavioral Ecology and Sociobiology* 29: 197–204.

Isaac, Glynn. 1978. "The food-sharing behavior of protohuman hominoids." *Scientific American* 238: 90–108.

Johnson, Randall H. 1982. "Food-sharing behavior in captive Amazon River dolphins (*Inia geoffrensis*)." *Cetology* 43: 1–2.

Kaplan, Hillard, Hill, Kim, Cadeliña, Rowe V., Hayden, Brian, Hyndman, David Charles, Preston, Richard J., Smith, Eric Alden, Stuart David E. and Yesner David R. 1985. "Food sharing among Ache foragers: Tests of explanatory hypotheses." *Current Anthropology* 26: 223–246.

Mech, David L. 1970. *The wolf: The ecology and behavior of an endangered species*. New York: Natural History Press.

Mitani, John C. and Watts, David P. 1999. "Demographic influences on the hunting behavior of chimpanzees." *American Journal of Physical Anthropology* 109: 439–454.

Mitani, John C. and Watts, David P. 2001. "Why do chimpanzees hunt and share meat?" *Animal Behaviour* 61: 915–924.

Parker, Sarah Taylor and Gibson, Karen R. 1979. "A developmental model for the evolution of language and intelligence in early hominids." *The Behavioral and Brain Science* 2: 367–408.

Perry, Susan and Rose, Lisa. 1994. "Begging and transfer of coati meat by white-faced capuchin monkeys, *Cebus capucinus*." *Primates* 35: 409–415.

Pika, Simone. 2008a. "Gestures of apes and pre-linguistic human children: Similar or different?" *First Language* 28: 116–140.

Pika, Simone. 2008b. "What is the nature of the gestural communication of great apes?" In *The Shared Mind*, Jordan Zlatev, Tim Racine, Chris Sinha and Esa Itkonen (eds), 165–186. Amsterdam/Philadelphia: John Benjamins Publishing Company.

Pika, Simone. 2009. "Our grooming cousins: Providing the link to declarative signalling?" *Studies in Communication Sciences* 9: 73–102.

Pika, Simone and Mitani, John C. (forthcoming). "The impact of communication: Success and failure in meat sharing events of chimpanzees in the wild."

Pika, Simone and Mitani, John C. 2006. "Referential gesturing in wild chimpanzees (*Pan troglodytes*)." *Current Biology* 16: 191–192.

Pika, Simone and Mitani, John C. 2008. "Referential gestures in chimpanzees in the wild: Precursors to symbolic communication?" In *The evolution of language: Proceedings of the 7th*

International Conference (Evolang7), Andrew D. M. Smith, Kenny Smith and Ramón Ferrer i Cancho (eds), 478–479. London: World Scientific Publishing.
Rose, Lisa and Marshall, Fiona. 1996. "Meat eating, hominid sociality, and home bases revisited." *Current Anthropology* 37: 307–338.
Sacks, Harvey, Schegloff, Emanuel A. and Jefferson, Gail. 1974. "A simplest systematics for the organization of turn-taking in conversation." *Language* 50: 696–735.
Schaller, George B. 1972. *The Serengeti Lion: A study of predator-prey relations*. Chicago, London: University of Chicago Press.
Schegloff, Emanuel A. 1984. "On some gestures' relation to talk." In *Structures of Social Action: Studies In Conversation Analysis*, John M. Atkinson and John C. Heritage (eds). Cambridge: Cambridge University Press.
Schegloff, Emanuel A. 2007. *Sequence Organization*. Cambridge: Cambridge University Press.
Scheid, Christelle, Schmidt, Judith and Noe, Ronald. 2008. "Distinct patterns of food offering and co-feeding in rooks." *Animal Behaviour* 76: 1701–1707.
Scott, Nicole M. and Pika, Simone. 2012. "A Call for Conformity: Gesture studies in human and non-human primates." In *Developments in Primate Gesture Research*, Simone Pika and Katja Liebal (eds), 145–161. Amsterdam/Philadelphia: John Benjamins Publishing Company.
Sidnell, Jack. 2010. *Conversation analysis: An introduction*. Chichester: Wiley-Blackwell.
Stevens, Jeffrey R. 2004. "The selfish nature of generosity: Harassment and food sharing in primates." *Proceedings of the Royal Society of London, Series B* 271: 451–456.
Stevens, Jeffrey R. and Gilby, Ian C. 2004. "A conceptual framework for nonkin food sharing: Timing and currency of benefits." *Animal Behaviour* 67: 603–614.
Stirrat, Michael, Gumert, Michael and Perrett, David. 2011. "The effect of attractiveness on food sharing preferences in human mating markets." *Evolutionary Psychology* 9: 79–91.
Struhsacker, Tom T. 1997. *Ecology of an African rain forest*. Gainesville, Florida: University Presses of Florida.
Takahata, Yukio, Hasegawa, Toshikazu and Nishida, Toshisada. 1984. "Chimpanzee predation in the Mahale Mountains from August 1979 to May 1982." *International Journal of Primatology* 5: 213–233.
Teleki, Geza. 1973. *The predatory behavior of wild chimpanzees*. Brunswick: Bucknell University Press.
ten Have, Paul. 2007. *Doing Conversation Analysis*. London: Sage.
Vahed, Karim. 1998. "The function of nuptial feeding in insects: A review of empirical studies." *Biological Reviews* 73: 43–78.
Van Hooff, Jan A. R. A. M. 1973. "A structural analysis of the social behaviour of a semi-captive group of chimpanzees." In *Social Communication and Movement, Studies of Interaction and Expression in Man and Chimpanzee*, Mario von Cranach and Ian Vine (eds), 75–162. London: Academic Press.
Van Lawick-Goodall, Jane. 1968. "A preliminary report on expressive movements and communication in the Gombe stream chimpanzees." In *Primates. Studies in Adaptation and Variability*, Phyllis C. Jay (ed.), 313–374. New York: Holt, Rinehart, and Winston.
Wilkinson, G.S. 1992. "Communal nursing in the evening bat, *Nycticeius humeralis*." *Behavioral Ecology and Sociobiology* 31: 225–235.
Wilson, Edward O. 1975. *Sociobiology*. Cambridge: Belknap Press/Harvard University Press.
Winterhalder, Bruce. 1996. "Social foraging and the behavioral ecology of intragroup resource transfers." *Evolutionary Anthropology* 5: 46–57.

Wrangham, Richard W. 1975. "*The behavioural ecology of chimpanzees in the Gombe National Park, Tanzania.*" Unpublished Ph.D. thesis, Cambridge University, Cambridge.

Wrangham, Richard W., van Zinnicq, E. and Riss, Bergmann. 1990. "Rates of predation on mammals by Gombe chimpanzees, 1972–1975." *Primates* 31: 157–170.

Wright, Jonathan and Leonard, Marty L. (eds). 2002. *The evolution of nestling begging: Competition, cooperation and communication.* Dordrecht: Kluwer.

Hands and faces

Linking human language and non-human primate communication

Sherman Wilcox
Department of Linguistics, University of New Mexico

This chapter investigates the linguistic links between gesture and language, relying on data from signed languages. Data focuses on the developmental routes by which non-linguistic gestures become incorporated into the linguistic system of signed languages. Evidence is presented for two routes: one leading from gesture to word to grammatical morpheme, and a second leading from gesture to prosody to grammatical marker. These two routes emerge because of the semiotic characteristics of hands and faces. One application of this work is to inform research on non-human primate gesture. The goal is to pose new questions about the relation between gesture and language and to propose a new approach that primate gesture researchers may address when surveying the range of gestural forms.

1. Introduction

Questions of origin are always fascinating: where did the universe come from, where did life come from, where did humans come from, and where did language come from? The question seems especially difficult for cosmologists who have to figure out how something, the universe, came from nothing. For the rest of us, the questions are somewhat easier because we do not have to propose that something came from nothing. Life emerged from inorganic chemicals, humans have ancestral precursors, and unless language sprang fully formed from Jove's brow, it too has precursors (Hockett 1978).

One goal of science is to find precursors, to find connections. Without connections there can be no cause and effect, without cause and effect there can be no explanation, and without explanation we are left with mystery. As the anthropologist Matt Cartmill (1990: 188–89) has observed:

> To understand the origin of anything, we must have an overarching body of theory that governs both the thing itself and its precursors. Without such a body of theory, we have no way of linking the precursor to its successors, and we are

> left with an ineffable mystery, like the one that Chomsky ... and Lenneberg ... have always insisted must lie at the origin of syntax. ... Our aim as scientists must be to explain as much as possible: to find animal analogs for human traits and seek, as students of primate behavior have long sought, common patterns of adaptation underlying those analogs.

Linguists have not always been sympathetic to overarching theories that link language with its non-language precursors. Some linguists in fact do propose "something out of nothing" theories. Chomsky (1972: 70) for example, declares, "There seems to be no substance to the view that human language is simply a more complex instance of something to be found elsewhere in the animal world." Concerning any possible relation between language, animal communication, and other forms of human communication such as gesture, he concludes that,

> The examples of animal communication that have been examined to date do share many of the properties of human gestural systems, and it might be reasonable to explore the possibility of direct connection in this case. But human language, it appears, is based on entirely different principles.

If we follow Chomsky the conclusion is two-fold: animal communication is distinct from human language, and, although there is a possible connection between human gestural and animal communication systems, human gestural communication is also distinct from human language. According to this view, there are no precursors to language, and no links between human gesture and human language.

In this chapter, I will explore two alternatives to Chomsky's view: First, I will demonstrate that non-linguistic gesture does have a direct connection with human language in the case of the incorporation of gesture into signed languages. Second, I will suggest an overarching theory that can inform the exploration and comparison of human gesture, human language, and non-human primate communication.

In order to begin constructing an overarching theory linking language with its precursors, we need to adopt three foundational frameworks: (1) Language as gesture, (2) Dynamic system theory, and (3) Embodied grammar based on embodied cognition. The first two will be outlined below. The third plays a lesser role in the issues under consideration here, but is discussed in more detail in other writings (Armstrong & Wilcox 2007; Wilcox 2004).

1.1 Language as gesture

The first framework unites language, both spoken and signed languages, as ultimately gestural in its nature. Spoken and signed languages are produced by articulatory movements – of the vocal tract of speakers, and of the hands and faces of signers – which in turn produce perceptible signals. This framework can be traced to cognitive scientists such as Ulrich Neisser (1967: 156), who noted that,

> to speak is to make finely controlled movements in certain parts of your body, with the result that information about these movements is broadcast to the environment. For this reason the movements of speech are sometimes called articulatory gestures.

The framework is also adopted by those who study speech as the production and perception of articulatory gestures (e.g. Browman & Goldstein 1985, 1989, 1992). By studying this dynamic organization of speech as articulatory gestures, Kelso and his colleagues have come to the conclusion that, "words are not simply strings of individual gestures, produced one after the other; rather, each is a particular pattern of gestures, orchestrated appropriately in time and space" (Kelso, Saltzman & Tuller 1986: 31).

Although little research has been conducted applying this articulatory gesture approach to signed languages, it is clear that signed words are also patterns of articulatory gestures, orchestrated appropriately in time and space. Thus, rather than unifying spoken and signed languages by adopting an abstractionist solution, which abstracts away from the disparate physical systems and proposes an "amodal" linguistic system that is blind to its actual production, the articulatory gesture approach embraces an embodied solution that grounds both spoken and signed languages in their physical manifestation as the production of perceptible movements.

It should be noted that by adopting this framework another unification is made, one that is quite important for understanding the precursors of language. The framework also unifies language and gesture: If words, signed or spoken, are patterns of articulatory gestures orchestrated in time and space, then what we commonly call 'gestures' are also composed of articulatory gestures orchestrated in time and space. This fact applies both to instrumental gestures such as grasping, whether produced by animals or humans, and to symbolic gestures such as the co-speech gestures widely studied by gesture researchers (Cienki & Müller 2008; Kendon 1995, 2004; McNeill 1992, 2000, 2005). This framework also enables the scientist to study the gestural communication systems of human and non-human primates under the same theoretical model.

An increasing number of researchers point to gesture as an evolutionary precursor to language (Armstrong & Wilcox 2007; Armstrong, Stokoe & Wilcox 1995; Corballis 1999; Gentilucci 2006; Hewes 1992; Morillon et al. 2010; Rizzolatti & Arbib 1998). Evidence suggests that gesture also plays a critical role in the ontogenetic development of language (Blake 2000; Capirci, Caselli, Iverson, Pizzuto & Volterra 2002).

Finally, research from cognitive neuroscience also supports the linking of gesture and language. Edelman (1989) for example, presents a comprehensive non-Cartesian model of the evolutionary emergence of language as an epigenetic phenomenon. Edelman's model relies on neuronal group selection to explain how the brain self-organizes. A key feature of his model is the role that muscle synergies play in the development of both concept formation and presyntax.

According to the neuroscientist Rudolfo Llinás (2001), muscle synergies are defined as collections of muscle fibers that are successful in achieving some purpose. This definition evokes the same general notion of gesture that I apply, which is derived from dynamic systems theory: "A gesture is a functional unit, an equivalence class of coordinated movements that achieve some end" (Studdert-Kennedy 1987:77).

1.2 Dynamic systems

Chomsky is also wrong that complex systems cannot emerge from simpler systems. Nature is replete with examples of more complex structure emerging from simpler systems (Kelso 1995). The study of such emergent systems is the domain of dynamic systems theory, the second framework that must be adopted in order to link language and its precursors.

A prototype example of such emergence is the so-called Belousov-Zhabotinskii chemical reaction. As a chemist mixes a few chemicals in a dish, a pattern of concentric, circular rings appears; the rings expand, collide, and develop spirals, organizing into unique patterns, which eventually decay and die out. Thelen and Smith (1994:46) describe the event:

> As chemists increase the flow of the reactants in a smooth and continuous manner, they find remarkable and curious results. Although they increase the reactants gradually, the oscillation periods increase by precise doublings of the original period. Then, at a critical flow rate, the reaction seems to go wild; the ions fluctuate erratically and seemingly randomly.

What has happened? A gradual and continuous change in one variable has led to a nonlinearity in another. As Thelen and Smith (1994:49) describe it, the molecules have cooperated as part of a dynamic system "to produce complexity from the most simple of precursors, to generate patterns without a designer, and to set time ticking off without a watchmaker."

Dynamic systems theory has been applied to the study of movement development (Thelen 1996; Thelen et al. 1993), speech production (Browman & Goldstein 1985, 1989, 1992; Kelso et al. 1986), and cognition (Smith 2005a,

2005b; Thelen et al. 1994). Linguists have outlined a general approach to language based on dynamic systems theory (Elman 1998; Port & van Gelder 1998; Van Geert 1991).

The same dynamic principles that underlie the emergence of complex systems across chemical, biological, cognitive, and motor domains can be applied to the emergence of language as a complex system from simpler communication systems, without the need for a divine watchmaker and without ineffable mysteries.

2. Gesture to language

Although not specifically addressing gesture, Haiman (1998: 128) proposes that ritualization can account for the emergence of language from non-language:

> The ... evolution of language from originally instrumental action to symbolic is plausible: elsewhere in the animal kingdom, semanticization or emancipation occurred wherever originally instrumental acts were modified and stylized to produce signals.

By invoking the notion of ritualization, Haiman is pointing to a thoroughly studied linguistic process called grammaticization by which the more grammatical elements of a language, such as grammatical morphemes and markers, gradually develop from lexical morphemes (Bybee, Perkins & Pagliuca 1994).

In previous research, I have documented how non-linguistic gestures follow two developmental routes as they become a part of the linguistic system of a signed language (Armstrong & Wilcox 2007; Wilcox 2004, 2007). The first route begins with a manual gesture, which enters a signed language first as a lexical sign and then develops through grammaticalization into a grammatical morpheme. The second route proceeds along a distinctly different path. The source is not the manual gesture itself; rather, it is the way that a manual gesture is produced, as well as various facial, mouth, and eye gestures that may accompany a manual gesture or sign. Upon entering the linguistic system, these manner of movement and facial gestures follow a developmental path from prosody/intonation to grammatical markers.

In the sections to follow, I will extend this notion to the gesture-language connection and explore the process by which grammar emerges from gesture in signed languages. This is important, because it demonstrates that the cognitive processes and mechanisms known to be operating in the development of grammar across languages can be extended to account for how non-linguistic gesture becomes language. The two routes emerge because of different properties of gesture, which lead to two distinct grammaticization paths from gesture to language. By explicitly identifying these two paths, I hope to open the possibility that primatologists can more systematically explore the possible link between gesture and language.

2.1 The first route

Two sources of evidence for the developmental path leading from gesture to lexical morpheme to grammatical morpheme are presented here: futures and markers of epistemic modality. One example of grammaticization in action is the development of future markers. Data from a cross-section of the world's spoken languages demonstrate that there are three common sources for future markers: desire, obligation, and movement verb constructions (Bybee et al. 1994).

Using a corpus of historical as well as modern conversational data, Shaffer (2000) and Janzen and Shaffer (2002) have demonstrated that in American Sign Language (ASL) the grammatical morpheme used to mark future (see Figure 1a) developed from a lexical morpheme 'go' or 'depart' (see Figure 1b).

Figure 1a. Old ASL FUTURE

Figure 1b. Old ASL DEPART

The gestural source of the future morpheme is described by de Jorio (1832) as produced with the palm of the hand open and held edgewise, and moved upwards several times. Morris and his colleagues (Morris, Collett, Marsh & O'Shaughnessy 1979) identify this as a gesture still in use among hearing people in the Mediterranean region to signal departure-demand and departure-description (see Figure 2).

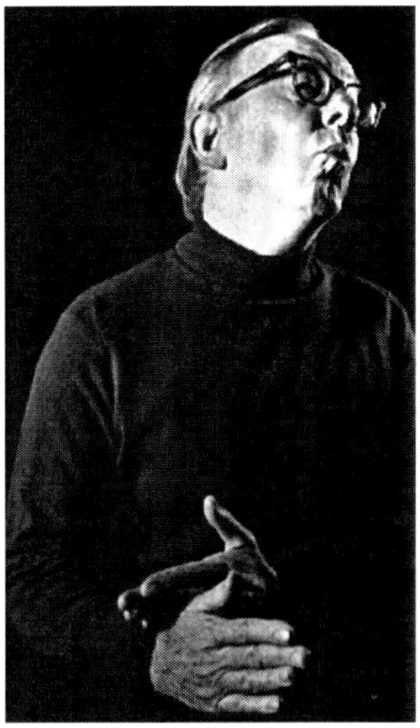

Figure 2. Gesture meaning DEPART (from Wylie & Stafford 1977)

The gesture appears in French Sign Language (LSF) of the 19th century as the lexical morpheme PARTIR 'depart'.

Other evidence for the grammaticalization of gesture following the first route comes from ASL and Catalan Sign Language (LSC). ASL modals of possibility have been shown to emerge from gestural sources expressing physical strength, which are incorporated into ASL as lexical signs indicating strength and then grammaticalize to modal forms (Wilcox & Wilcox 1995). In LSC, gestural forms indicating the eyes and visual perception, bright light, the nose and the sense of smell, and physical, facial appearance enter the linguistic system as lexical morphemes and develop into grammatical morphemes (Wilcox 2004).

A final example comes from the expression of epistemic modality in Italian Sign Language (LIS). Epistemic modality is an expression of the speaker's subjective evaluation of the likelihood of the proposition of a sentence being true. The LIS sign 'impossible' (see Figure 3) expresses epistemic modality. The sign appears to have a gestural source in the Roman Catholic benediction gesture, which in turn can be traced to an ancient Roman oratorical gesture that has been dubbed "the speaking hand" (Barasch 1987). The Roman "speaking hand" gesture is made with the thumb, index, and middle fingers extended, the ring and pinky fingers being closed into the palm; the gesturing hand is extended outward. The developmental route from its use in Roman oratory to liturgical settings, and from there to the LIS sign meaning 'impossible', has been described in detail (Wilcox 2009).

Figure 3. LIS sign IMPOSSIBLE (from Radutzky 1992)

The benediction gesture also became associated with death. This link is motivated by a natural metonymic contiguity, since the blessing spoken over a dead person is commonly accompanied with the benediction gesture, as seen in Giotto's "The Raising of Lazarus."

The association with death extended beyond the liturgical context, and at the same time took on an extended metaphorical meaning. De Jorio (1832: 281–283) describes a gesture he calls *Morte ("Death")*:

> The sign of the cross is made in the air with an extended hand. This gesture, that is also frequently done with just the index and middle fingers extended, can be used to mean physical death, but it can also refer to moral or political death, since it can be said of someone that he ceases to exist, as far as society is concerned or so far as the estimation of others is concerned In this case, the meaning of the gesture changes, not so much according to the facial expression that accompanies it, as it does according to what the interlocutors take the topic of the discourse to be. In fact, if the benediction is accompanied by a dejected and pained facial expression, and one deals with someone who is gravely ill, physical death will be understood. If the person being talked about is someone who enjoys great favor or who has a very showy job, it will be understood that he no longer enjoys either the one or the other.

The metaphorical extension from physical to moral or political death sanctioned a further metonymic link into the semantic domain of hopelessness and despair (de Jorio 1832: 283):

> Lost hope, despairing of one's business. It is the same as saying 'I am lost, it is finished'. Accordingly, if the good outcome of some activity is under discussion,

> and someone wishes to say that it is finished, that there is no more hope, it will be sufficient to make a cross in the air, either with the hand or with only two fingers.

This metonymic association of the benediction gesture with death, as seen in the *Morte* gesture, is the likely proximal source of the LIS sign 'impossible'. If the *Morte* gesture crossed into the linguistic system of LIS, we would expect it to first appear in a more lexical sense. This is in fact the case; the LIS sign 'death' is formationally related to the benediction gesture (see Figure 4). The extension from this lexical meaning to the more grammatical epistemic modality sense requires a model that can account for the death, hopelessness, and epistemic modality as part of a coherent semantic domain. This model comes in the form of an idealized cognitive model of the forward evolution of a conceived reality.

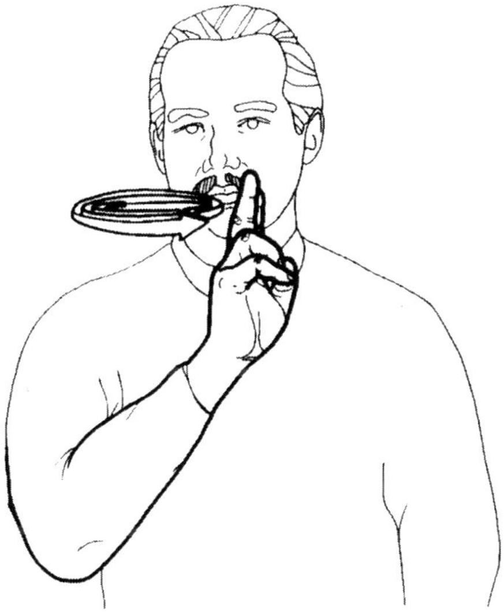

Figure 4. LIS sign DEATH (from Radutzky 1992)

Death, hopelessness, and despair are already in the semantic domain of modals. Givón (1984) points out that epistemic verbs include such notions as probability, possibility, uncertainty, doubt, hope, and fear. Modal verbs as a class evoke the conception of some associated activity encoded in the main verb. The subject of the modal verb is the locus of some kind of potency directed at the main verb, which drives reality forward in some way (Langacker 1991).

What emerges is an idealized cognitive model of modals in which a core concept is the likelihood of reality evolving in a certain way (Langacker 1991), whether the force that drives reality forward is a person's mental or physical ability, general conditions, the force of authority as evidenced in deontic senses such as *I can read Chinese* or *You must pay your taxes on time*, or the speaker's subjective assessment of the likelihood of reality evolving in some particular way, as in the epistemic sense in *If you didn't sleep at all last night, you must be tired* This view of modality is captured by Langacker in what he calls a **dynamic evolutionary model** (see Figure 5). In this model, reality is conceived as moving forward along a temporal axis, from the present reality into some projected future reality.

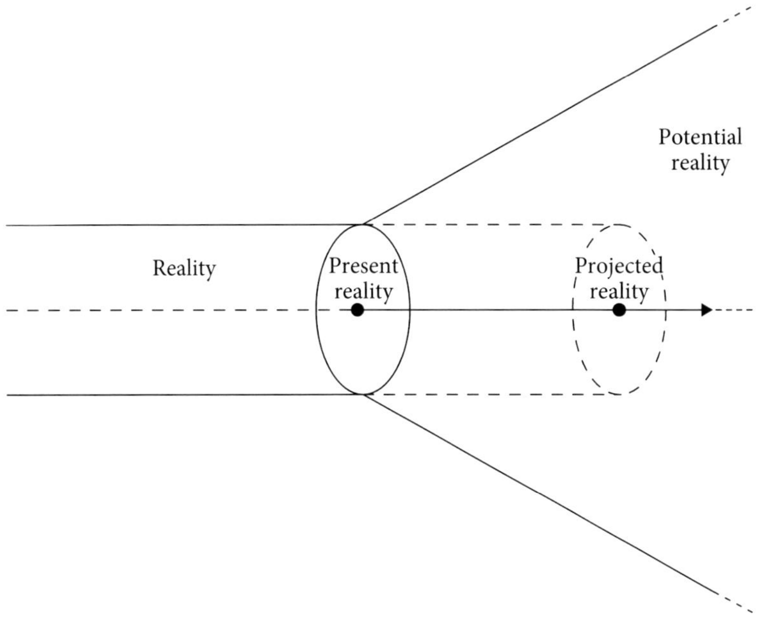

Figure 5. Dynamic evolutionary model (after Langacker 1991)

Under this view, then, we see that death metonymically leads to an association with hopelessness and despair because it precludes any possible future reality. Extending the semantics into the grammatical domain, the LIS sign 'impossible' expresses the speaker's evaluation of the impossibility of any projected or potential reality from coming into existence.

2.2 The second route

A gesture and its manner of movement, follow distinct developmental paths when the gesture enters the linguistic system. As we have just seen, manual gestures grammaticize along Route one, entering the linguistic system of signed languages first as lexical morphemes; they further develop into grammatical morphemes, functioning in such grammatical roles as the expression of future, venitives, and epistemic modality.

Movement consists of two aspects, what we might call its **what** and its **way**. When we perform an action such as grasping, lifting, hitting, or chopping, we perform both the **what**, the action itself, and the **way**, the way in which we are doing the action; the way we are grasping, lifting, hitting, or chopping. I call this second aspect of movement its manner of movement. While the **what** aspect of movement follows the developmental route described above as Route one, the *way* aspect follows Route two, which I will describe in this section. Of course, every gesture is performed with both a **what** aspect and a **way** aspect; it is in the linguistic development of these two aspects that they diverge, taking distinct developmental routes and expressing distinct grammatical functions.

Unlike Route one, in which the gestures entered the linguistic system as lexical morphemes, the gestures associated with Route two bypass the lexical stage, entering the linguistic system as prosody/intonation and then becoming grammatical markers (Wilcox 2007).

Upon entering the linguistic system, the manner of movement associated with a gesture appears first as prosody/intonation. The use of manner of movement to mark prosody and intonation in signed languages is well documented Friedman (1977) observed that signs articulated with emphatic stress are larger, tenser, faster, and with longer duration than unstressed signs. Noting that in spoken languages the primary cues for linguistic stress are increased duration, increased intensity, and changing the fundamental frequency, Wilbur and Schick (1987) have proposed that the correlates of linguistic stress in signed languages include increased duration and increased intensity. Markers of increased duration include larger movement, slower movement, repetition, and added movement. The indicators of increased intensity include the addition of non-manuals (for example, eye or mouth gestures), sharp boundaries between signs, higher articulation of signs in the signing space, increased tension of articulation, and a more forceful articulation. Reporting data from Israeli Sign Language, Nespor and Sandler (1999) found that the phonetic correlates of prosodic prominence included reduplication, a hold at the end of the prominent sign, and a pause after the last word of the phonological phrase. Once again, we find that the phonetic correlates to prosody in signed languages lie in manner of movement: reduplication, hold (stopping the movement), speed, and acceleration.

In ASL, in addition to expressing prosody, manner of movement is used to mark grammatical functions such as verb aspect (Klima & Bellugi 1979); to mark different adjectival predicate forms such as the distinction between sick, sickly, and prone to being sick (Klima & Bellugi 1979); and to indicate intensification along various axes. Variations in manner of movement also express distinctions of modal strength. For example, the distinction between weak obligation and strong obligation in ASL is marked by changes in manner of movement: weak obligation is indicated by weak, sometimes reduplicated movement, while strong obligation is marked by a single, forceful movement. These more grammatical functions of manner of movement, I suggest, are grammaticized functions of prosody/intonation, – a feature which had its pre-linguistic source in the way in which an action is made.

Manner of movement marks some of these same grammatical functions in other signed languages as well. Jarque (2006) describes the use of manner of movement in LSC to mark differences in modal strength and to distinguish deontic versus epistemic function. Pizzuto (1987) notes that temporal aspect can be expressed in LIS through systematic alterations of a verb's manner of movement. The "suddenness" of an action is marked by tense, fast, short movements (e.g. the distinction between 'to meet' and 'to suddenly/unexpectedly meet someone'). A verb produced using a large elliptical, slow movement specifies that an action is 'repeated over and over in time' or 'takes place repeatedly in time' (e.g. to constantly telephone or to always be on the telephone). Modal strength is also marked in LIS by weak or strong articulation of the movement.

Facial gestures play a significant role cross-linguistically in signed languages (Janzen 1999). In addition to expressing emotion, much the same way that tone of voice functions in spoken languages, facial gestures mark a variety of grammatical functions in many signed languages. These grammatical functions include the marking of interrogatives, topics, adverbials, conditionals, imperatives, and more. In an extensive study of interrogatives in 35 signed languages, Zeshan (2004) found that all used facial marking to indicate polar questions. Polar questions were marked by a cluster of facial gestures that included eyebrow raise, eyes wide open, eye contact with the addressee, head forward position, and forward body posture. In ASL, similar facial gestures are used to mark the if-clause in conditional expressions, and topic clauses in topic-comment expressions across a number of signed languages. In Spanish Sign Language, facial gestures both accompany the expression of grammatical modality and, in some case, function on their own to mark modal meanings (Lago 2006).

Clearly these linguistic facial markers have their source in non-linguistic facial gestures. Consider, for example, the cluster of facial gestures described above that are used to mark polar questions and other grammatical functions. A similar constellation of facial gestures is associated with surprise (Darwin 1872).

Although less studied than the grammaticalization of manner of movement, research, such as that reported here on the development of non-linguistic facial gestures into linguistic expressions of prosody-intonation, which then appear to grammaticize into grammatical markers, holds great promise for revealing the links between gesture and language.

3. Linking human language and non-human primate communication

Searching for the evolutionary precursors to human language is a bit like putting Humpty Dumpty back together again. The difference is that in the case of Humpty, the complexity of the task was orders of magnitude simpler. Before his fall, we knew what Humpty looked like. After his fall, the pieces may have been scattered, but they were still pieces of Humpty. All that the King's men had to do was collect up the pieces and assemble them together again.

The task is much more complex when it comes to linking human language and non-human primate communication in the search for the origins of language. This task depends not just on putting the pieces back together. In this case, we don't even know what the pieces are. Which theory of language should we choose? Which theory of cognition? Which theory of brain structure and function? What is the relation between cognition and language? What do we assume to be the relation between signed and spoken languages? Is language gestural? What is the relation between language and gesture? All of these questions are pieces of the puzzle, and all have different answers depending on which theory one accepts. Not just any linguistic theory, or brain model, will work. It is as if all the people in Humpty's village collected up and deposited into a pile what they thought were pieces of Humpty; some of them are indeed actual pieces of Mr. Dumpty, but others are just random pieces of potshards or other material that will never fit together to make a coherent Humpty. And, even after we have selected what we hope are the correct pieces, the correct theories and frameworks, we still have to put them together in the right way.

I have suggested that the critical pieces of the puzzle that can lead to an overarching theory of the emergence of human language are a framework that unites spoken and signed languages, along with gesture, as articulatory movements organized by dynamic systems. My hope is that such an overarching theory can inform the exploration and comparison of human gesture and language with non-human primate gestural communication (Call & Tomasello 2008; Liebal, Müller & Pika 2007).

I also hope that documenting the two routes that gesture takes when as it grammaticizes and becomes language can help primatologists more systematically to explore the links between gesture and language. It seems to me that many of the current attempts to achieve this have focused more on what I am here calling Route one, investigating, and in some cases comparing, manual gestures in non-human primates with human gestures. I would hope that analysis of manner movement as a feature of what I am calling Route two would also inform such analyses. We might ask, how does the **way** in which a gesture, whether manual, postural, or produced with some combination of limbs and body, impact the function and meaning of the gesture? How, for example, do animals distinguish between real fighting and play fighting? Is it perhaps some modification of the parameters involved in manner of movement? The question is important because play fighting can be seen as on the path to becoming symbolic.

Route two also holds great promise for comparing connections between human and non-human primate communication. The modulation of manner of movement has been described for cross-fostered chimpanzees learning ASL (e.g. Chalcraft & Gardner 2005). The use of facial gestures is described by Maestripieri (2005), Parr (2008; Parr, Waller & Fugate 2005; Parr, Waller & Vick 2007), and others (Pollick & de Waal 2007; Tanner & Byrne 1993; Zeller 1996). Müller (2007) has presented a comprehensive outline for comparative studies of gesture in human and non-human primates, setting out the conceptual and methodological grounds for such an approach in great detail. Her approach is, in my view, quite compatible with the framework proposed here.

References

Armstrong, David F. and Wilcox, Sherman. 2007. *The gestural origin of language.* Oxford, New York: Oxford University Press.

Armstrong, David F., Stokoe, William C. and Wilcox, Sherman. 1995. *Gesture and the nature of language.* Cambridge: Cambridge University Press.

Barasch, Moshe. 1987. *Giotto and the language of gesture.* Cambridge: Cambridge University Press.

Blake, Joanna. 2000. *Routes to child language: Evolutionary and developmental precursors.* Cambridge/New York: Cambridge University Press.

Brouland, Josephine. 1855. *Langage mimique: Spécimen d'un dictionaire des signes.* Washington, DC: Gallaudet Archives.

Browman, Catherine P. and Goldstein, Louis. 1985. "Dynamic modeling of phonetic structure." In *Phonetic linguistics.* Victoria A. Fromkin (ed), 35–54. NY: Academic Press.

Browman, Catherine P. and Goldstein, Louis. 1989. "Articulatory gestures a phonological units." *Phonology* 6: 201–251.

Browman, Catherine P. and Goldstein, Louis. 1992. "Articulatory phonology." *Phonetica* 49: 155–180.

Bybee, Joan, Perkins, Revere and Pagliuca, William. 1994. *The evolution of grammar: Tense, aspect, and modality in the languages of the world.* Chicago: University of Chicago Press.

Call, Josep and Tomasello, Michael. 2008. *Gestural communication of Apes and Monkeys.* Mahwah, New York: Lawrence Erlbaum Associates.

Capirci, Olga, Caselli, M. Cristina, Iverson, Jana M., Pizzuto, Elena A. and Volterra, Virginia. 2002. "Gesture and the nature of language in infancy: The role of gesture as a transitional device en route to two-word speech." In *The study of signed languages: Essays in honor of William C. Stokoe,* David F. Armstrong, Michael A. Karchmer and John V. Van Cleve (eds), 213–246. Washington, DC: Gallaudet University Press.

Cartmill, Matt. 1990. "Human uniqueness and theoretical content in paleoanthropology." *International Journal of Primatology* 11 (3): 173–192.
Chalcraft, Valerie J. and Gardner, Allan. 2005. "Cross-fostered chimpanzees modulate signs of American Sign Language." *Gesture* 5: 107–132.
Chomsky, Noam. 1972. *Language and mind.* NY: Harcourt Brace Jovanovich.
Cienki, Alan J. and Müller, Cornelia. 2008. *Metaphor and gesture.* Amsterdam/Philadelphia: John Benjamins Publishing Company.
Corballis, Michael C. 1999. "The gestural origins of language." *American Scientist* 87 (2): 138–145.
Darwin, Charles. 1872. *The expression of the emotions in man and animals.* London: Murray.
Edelman, Gerald. M. 1989. *The remembered present: A biological theory of consciousness.* New York: Basic Books.
Elman, Jeffrey L. 1998. Language as a dynamical system. In *Mind as motion: Explorations in the dynamics of cognition,* Robert F. Port and Tim van Gelder (eds), 195–225. Cambridge, MA: The MIT Press.
Friedman, Lynn A. 1977. *On the other hand: New perspectives on American sign language.* New York, NY: Academic Press.
Gentilucci, Maurizio. 2006. "From manual gesture to speech: A gradual transition." *Neuroscience and Biobehavioral Reviews* 30 (7): 949–960.
Givón, Talmy. 1984. *Syntax: A functional-typological introduction.* Amsterdam/Philadelphia: John Benjamins Publishing Company.
Haiman, John. 1998. "The metalinguistics of ordinary language: Some case studies." *Evolution of Communication* 2 (1): 117–135.
Hewes, Gordon. W. 1992. "Primate communication and the gestural origin of language." *Current Anthropology* 33 (1–2): 65–84.
Hockett, Charles. 1978. "In search of Jove's brow." *American Speech* 53: 243–315.
Janzen, Terry. 1999. "The grammaticization of topics in American Sign Language." *Studies in Language* 23: 271–306.
Janzen, Terry and Shaffer, Barbara. 2002. "Gesture as the substrate in the process of ASL grammaticization." In *Modality and structure in signed and spoken languages,* Richard Meier, David Quinto and Kearsy Cormier (eds), 199–223. Cambridge: Cambridge University Press.
Jarque, Maria J. 2006. "The expression of possibility in Catalan Sign Language: The sign PODER." In *5th International Conference of the Spanish Cognitive Linguistics Association (AELCO/SCOLA). Murcia, October 19–21 2006.*
de Jorio, Andrea. 1832. *Gesture in Naples and gesture in classical antiquity: A translation of la mimica degli antichi investigata nel gestire napoletano, gestural expression of the ancients in the light of neapolitan gesturing* (A. Kendon, Trans.). Bloomington, Indiana: Indiana University Press.
Kelso, J. A. Scott. 1995. *Dynamic patterns: The self-organization of brain and behavior.* Cambridge, MA: The MIT Press.
Kelso, J. A. Scott, Saltzman, Elliot L. and Tuller, Betty. 1986. "The dynamical perspective on speech production: Data and theory." *Journal of Phonetics* 14: 29–59.
Kendon, Adam. 1995. "Gestures as illocutionary and discourse structure markers in southern Italian conversation." *Journal of Pragmatics* 23 (3): 247–279.
Kendon, Adam. 2004. *Gesture: Visible action as utterance.* Cambridge/New York: Cambridge University Press.
Klima, Edward and Bellugi, Ursula. 1979. *The signs of language.* Cambridge, MA: Harvard University Press.

Lago, Silvia. I. 2006. *Uso del componente facial para la expresión de la modalidad en lengua de signos española.* Unpublished dissertation, Universidade de Vigo, Spain.

Langacker, Ronald W. 1991. *Foundations of cognitive grammar. Volume II, descriptive application.* Stanford: Stanford University Press.

Liebal, Katja, Müller, Cornelia and Pika, Simone. 2007. *Gestural communication in nonhuman and human primates.* Amsterdam/Philadelphia: John Benjamins Publishing Company.

Llinás, Rudolfo R. 2001. *I of the vortex: From neurons to self.* Cambridge, Mass: The MIT Press.

Maestripieri, Dario. 2005. "Gestural communication in three species of macaques (*Macaca mulatta, M. Nemestrina, M. Arctoides*): Use of signals in relation to dominance and social context." *Gesture* 5 (1/2): 57–73.

McNeill, David. 1992. *Hand and mind: What gestures reveal about thought.* Chicago: University of Chicago Press.

McNeill, David. 2000. *Language and gesture.* Cambridge: Cambridge University Press.

McNeill, David. 2005. *Gesture and thought.* Chicago: University of Chicago Press.

Morillon, Benjamin, Lehongre, Katia, Frackowiak, Richard S. J., Ducorps, Antoine, Kleinschmidt, Andreas, Poeppel, David, et al. 2010. "Neurophysiological origin of human brain asymmetry for speech and language." *Proceedings of the National Academy of Sciences of the United States of America* 107 (43): 186–188.

Morris, Desmond, Collett, Peter, Marsh, Peter and O'Shaughnessy, Marie. 1979. *Gestures: Their origin and distribution.* New York: Stein and Day.

Müller, Cornelia. 2007. "Gestures in human and nonhuman primates: Why we need a comparative view." In *Gestural communication in nonhuman and human primates,* Katja Liebal, Cornelia, Müller and Simone Pika (eds), 237–60. Amsterdam/Philadelphia: John Benjamins Publishing Company.

Neisser, Ulrich. 1967. *Cognitive Psychology.* NY: Appleton-Century-Crofts.

Nespor, Marian and Sandler, Wendy. 1999. "Prosody in Israeli Sign Language." *Language and Speech* 42 (2–3): 143–176.

Parr, Lisa A., Waller, Bridget M. and Fugate, Jennifer. 2005. "Emotional communication in primates: Implications for neurobiology." *Current Opinions in Neurobiology* 15 (6): 716–720.

Parr, Lisa A., Waller, Bridget M. and Vick, Sarah J. 2007. "New developments in understanding emotional facial signals in chimpanzees." *Current Directions in Psychological Science* 16 (3): 117–122.

Pizzuto, Elena. 1987. "Aspetti morfosintattici." In *La lingua italiana dei segni – la comunicazione visivo-gestuale dei sordi,* Virginia Volterra (ed), 179–209. Bologna: Il Mulino.

Pollick, Amy S. and de Waal, Frans B. 2007. "Ape gestures and language evolution." *Publication of the National Academy of Science of the United States of America* 104 (19): 8184–8189.

Port, Robert F. and van Gelder, Tim. 1998. *Mind as motion: Explorations in the dynamics of cognition.* Boston, MA: The MIT Press.

Radutzky, Eelena. 1992. *Dizionario bilingue elementare della lingua italiana dei segni.* Rome: Edizioni Kapp.

Rizzolatti, Giacomo and Arbib, Michael A. 1998. "Language within our grasp." *Trends in Neuroscience* 21 (5): 188–194.

Shaffer, Barbara. 2000. *A syntactic, pragmatic analysis of the expression of necessity and possibility in American sign language.* Unpublished dissertation, University of New Mexico, USA.

Smith, Linda B. 2005a. "Action alters shape categories." *Cognitive Science* 29 (4): 665–679.

Smith, Linda B. 2005b. "Cognition as a dynamic system: Principles from embodiment." *Developmental Review* 25 (3–4): 278–298.

Studdert-Kennedy, Michael. 1987. "The phoneme as a perceptuomotor structure." In *Language perception and production: Relationships between listening, speaking, reading, and writing*, Alan Allport (ed), 67–84. London: Academic Press.

Tanner, Joanne E. and Byrne, Richard W. 1993. "Concealing facial evidence of mood: Perspective-taking in a captive gorilla?" *Primates* 34 (4): 451–457.

Thelen, Esther. (1996). "Motor development: A new synthesis." In *Annual progress in child psychiatry and child development*, Margaret E. Herzog and Ellen A. Farber (eds), 32–66. Philadelphia: Brunner-Routledge.

Thelen, Esther, Corbetta, Daniela, Kamm, Kathi, Spencer, John P., Schneider, Klaus and Zernicke, Ronald F. 1993. "The transition to reaching: Mapping intention and intrinsic dynamics." *Child Development* 64 (4): 1058–1098.

Thelen, Esther and Smith, Linda B. 1994. *A dynamic systems approach to the development of cognition and action*. Cambridge, MA: The MIT Press.

Van Geert, Paul. (1991). "A dynamic systems model of cognitive and language growth." *Psychological Review* 98 (1): 3–53.

Wilbur, Ronnie B. and Schick, Brenda. 1987. "The effects of linguistic stress on ASL signs." *Language and Speech* 30 (4): 301–323.

Wilcox, Sherman. 2004. "Gesture and language: Cross-Linguistic and historical data from signed languages." *Gesture* 4 (1): 43–75.

Wilcox, Sherman. 2007. "Routes from gesture to language." In *Verbal and signed languages: Comparing structures, constructs and methodologies*, Elena Pizzuto Antinori, Paola Pietrandrea and Raffaele Simone (eds), 107–31. Berlin: Mouton de Gruyter.

Wilcox, Sherman. 2009. "Symbol and symptom: Routes from gesture to signed language." *Annual Review of Cognitive Linguistics* 7: 89–110.

Wilcox, S. and Wilcox, P. (1995). "The gestural expression of modality in American Sign Language." In Joan L. Bybee and Suzanne Fleischman (eds), *Modality in grammar and discourse*, 135–162. Amsterdam/Philadelphia: John Benjamins Publishing Company.

Wylie, Lawrence. 1977. *Beaux gestes: A guide to French body talk*. Cambridge: The Undergraduate Press.

Zeller, Anne C. 1996. The interplay of kinship organisation and facial communication in the macaques. In *Evolution and ecology of macaque societies*, John E. Fa and Donald G. Lindburg (eds), 527–550. New York: Cambridge University Press.

Zeshan, Ulrike. 2004. Interrogative constructions in signed languages: Crosslinguistic perspectives. *Language* 80 (1): 7–39.

Where next?

Liebal Katja[1,2,3] & Pika Simone[4]
[1]Freie Universität Berlin, Cluster Languages of Emotion, Department of Education and Psychology, Evolutionary Psychology, Habelschwerdter Allee 45, 14195 Berlin, Germany / [2]University of Portsmouth, Department of Psychology, King Henry 1st Street, PO1 2DY Portsmouth, UK / [3]Max Planck Institute for Evolutionary Anthropology, Department of Developmental and Comparative Psychology, Deutscher Platz 6, Leipzig, Germany / [4]Humboldt Research Group "Comparative Gestural Signalling", Max Planck Institute for Ornithology, Eberhard-Gwinner-Strasse, Geesehouse, 82319 Seewiesen, Germany

Since 2007, when we published our first book on *Gestural communication in non-human and human primates* as a result of an interdisciplinary workshop that brought together experts in very diverse fields of gesture research (Liebal, Müller & Pika 2007), the number of publications reporting new and fascinating results about research into gestural communication of primates has impressively increased. The current issue can of course only provide a glimpse on this exciting and vibrant field of research, but it hopefully demonstrates that there are some new and interesting avenues and developments.

The aim of this chapter is to identify foci of ongoing research as well as blind spots, such as remaining questions and unsolved debates that need to be addressed in future research into gestural communication of primates. Most of the contributions to this volume concern gesture use in non-human primates. This is not to suggest that there is no exciting research going on with regards to investigating gesture acquisition in pre-verbal children or in speech-accompanying gestures. However, gesture research in non-human primates is currently expanding at much higher rate with many more researchers now working on a variety of species in the framework of language evolution and in hope of discovering similarities between gestures used by humans and other primates to shed light on the evolution of human language.

1. Phylogenetic origins of gestural communication

The current chapter by van Hooff (this volume) nicely summarizes theories on language evolution that build on evidence of research on non-human primate species. The idea that gestures played an important role in the evolution of human language is already mentioned in the work of Quintilian in Roman Antiquity (Russell 2001). However, only in the last few decades, primate gestural communication has been investigated within an evolutionary comparative framework and by using a more psychological approach focusing on individual variability of gestural repertoires and their flexible use across different contexts. According to van Hooff (this volume: 15), "... inferences about evolutionary relationships should rest on broad comparative structural, motivational and functional analyses...". However, gestures in non-human primates have overwhelmingly been studied in great apes only, with the majority of research conducted on captive groups (Slocombe Waller & Liebal 2011). We still know very little about great apes' gestural communication in wild populations and even less about monkeys. However, as the current volume indicates, this is changing: Pika & Mitani (2006; 2008) were the first to investigate the referential use of gestures by chimpanzees in the wild. Subsequently, Hobaiter, Genty and Byrne described the gestural repertoires of chimpanzees and gorillas in their natural environments (Genty, Breuer, Hobaiter & Byrne 2009; Hobaiter & Byrne 2011), and investigated their use and combinations of gestures (Hobaiter & Byrne 2011b, 2012). For rehabilitant orangutans, Russon and Andrews (2011) reported that they even use pantomime as elaborations of failed requests. According to the authors, this included re-enacting of past events, declaratives, but also deceptions, and several features of language such as productivity and compositionality. Concerning monkeys, Laidre (2012, 2005, 2008, 2011) extensively published on gesture use by mandrills in both captive and wild populations. Still, to approach the question of the phylogenetic origins of primate communication and to identify the role of gestures in such evolutionary scenarios, much more evidence from monkeys and apes in both wild and captive settings is necessary to understand the richness and flexibility of their gestural communication.

2. Multimodal approach

This is tightly linked to another gap of knowledge, since current research on primate communication suffers from the lack of a multimodal approach. If we want to understand the factors driving the evolution of complex communicative systems, we need to consider more than one modality and study gestures in concert with facial expressions and vocalizations. Gestures do not occur in isolation, they are accompanied by facial expressions, shifts in body postures, or vocalizations (van Hooff this volume). A recent review of the literature on primate communication covering the last 60 years demonstrated that from a total of 553 studies on primate communication, only 5% considered more than one modality and studied them in an integrated way (Slocombe, Waller & Liebal 2011). However, a gesture combined with a facial expression might be used for a different function than when this gesture is used singly. Thus, it is essential to study different communicative means in concert if we want to achieve an overall picture of primate communication, which is even more important if we want to shed light on the evolution of human language.

3. Ontogenetic trajectories

In humans, children use gestures before they start to speak (Bates, Benigni, Bretherton, Camaioni & Volterra 1979) and even in adults, gestures are closely intertwined with language, either spoken or signed (Kendon 2000; McNeill 1992; Sandler 2009). Thus, language does not simply replace gestures – hand movements pave the way for language acquisition in pre-verbal children (Goldin-Meadow 2002, 2003). Furthermore, they convey important information in addition to speech itself (Mehrabian 1981). After the seminal work by Bates and colleagues (1976; 1979), Camaioni (1993), and Butterworth (2003), who studied the role of gestures and POINTING in particular in the framework of intentional communication and language acquisition and in concert with the underlying cognitive skills in human children, more recent work centres particularly around the social contexts and motives underlying the use of POINTING gestures in young children (e.g. Franco & Butterworth 1996; Liebal, Behne, Carpenter & Tomasello 2009; Liszkowski, Carpenter, Striano & Tomasello 2006). The cross-cultural variability of gesture use in children, which so far has been largely neglected in the literature (but see Blake, Vitale, Osborne & Olshansky 2005), is now also studied in other than Western cultures to understand how culture and socialization shape the way children communicate (Liszkowski, Brown, Callaghan, Takada & De Vos in press; Salomo & Liszkowski 2010).

How gestures other than POINTING emerge during ontogeny is, however, still not extensively studied in children, and we know even less about gestural ontogeny in non-human primates. There are studies that describe the gesture use of single or few individuals in a very detailed and often qualitative way or do not focus on gestures specifically, but refer to communicative behaviors (Chevalier-Skolnikoff 1973; Gómez 1990; Plooij 1979). Others focus on gestures that occur in a very specific context, such as BEGGING gestures used by infant orangutans to obtain food from their mothers (Bard 1992). A comprehensive investigation that included monkeys, gibbons, and great apes and followed their gestural communication over an extended period of time is largely unknown to the scientific community since it is published in Russian (Grigor'eva & Deryagina, 1987). Tomasello, Call and colleagues used a more psychological approach to gestural communication and covered almost two decades by their repeated observations of two groups of captive chimpanzees (Call & Tomasello 2007; Tomasello, Call, Nagell, Olguin & Carpenter, 1994; Tomasello, Call, Warren, Frost, Carpenter & Nagell 1997; Tomasello, George, Kruger Farrar & Evans 1985; Tomasello, Gust & Frost 1989). However, although some of the individuals were observed at different ages, the study in general used a cross-sectional design and thus did not follow most of the individuals over an extended period of time. It was found that chimpanzees do most likely not acquire their gestures by social learning (e.g. imitating the behavior of other group members), but by an individual learning process called ontogenetic ritualization. In ontogenetic ritualization, repeated interactions between dayds result in a high degree of variability between individual gestural repertoires (see also Marentette & Nicoladis this volume; Pika & Liebal this volume; Tempelman, Tanner & King this volume). In contrast, Genty and colleagues (2009) argue that this high variability is possible due to the fact that observations were not conducted long enough to capture the individuals' complete repertoires. They therefore conclude from their research on gorillas that their gestures and possibly those of other great apes are largely species-typical and thus genetically determined.

These inconsistencies between studies might be caused by differing criteria for defining a gesture, but they also demonstrate that more longitudinal studies are needed that follow several individuals over extended periods of time to investigate the mechanisms underlying gesture acquisition. Of course, such studies are difficult to conduct – both in terms of availability of subjects, logistical difficulties, and differing settings between zoos, as well as constraints on time. Schneider and colleagues (2011, in press) studied 25 individuals of all four great apes species and followed them over a period up to 20 months of their lives to determine the onset of gestural communication. Interestingly, orang-utans only started gesturing at around 15 months of age, thus at least four months later than bonobos, chimpanzees, and gorillas. Furthermore, a comparison of the gestural repertoires of infants and their mothers in both bonobos and chimpanzees suggests that infants of different groups share a larger repertoire than infants and their mothers. This suggests that infants very likely do not learn their gestures by observing their mothers, but that both a genetic component as well as the shaping by individual learning in interactions with others play a role for gesture acquisition in great apes. However, this data set is still too limited to actually answer the question how exactly non-human primates acquire their gestures and learn to use them appropriately.

4. Dynamic approach

Gestures or primate communication in general are often studied by using a rather static approach, that is looking at communication in a way that one individual (sender) uses a particular signal to get another individual (recipient) to respond. However, although it enables the observer to rather easily categorize different components of the information transfer, this approach ignores the fact that communication is not a simple one-way-street. Many factors may influence the way information is conveyed, such as the kind of environment, the identity of the interacting partners, as well as previous interactions of these two individuals, and so on. Therefore King and Shanker (2003) suggested studying gestural communication as a "dynamic dance" using a more qualitative approach. In the last few years, an increasing number of studies has been presented at different conferences, using a micro-analysis to investigate the development of non-human primate gestures (Halina & Rossano 2010; Johnson, Zastrow & Halina 2011), or their structural properties (Liebal, Bressem & Müller 2010). A new approach applied to study gestural communication of non-human primates is suggested by Wilkinson, Leudar and Pika (this volume). In this paper, a micro-analytic approach based on conversation analysis methods was applied for the first time to analyze interactions of wild chimpanzees in the context of food sharing. To apply such a method usually used to analyze spoken interactions to a non-linguistic species seems to be a paradox only at the first glance, since this very detailed approach might help to disentangle how having language actually shapes the way humans interact with each other. In other words, looking at the nonverbal behavior of non-human primates with a toll developed for human language might open very different perspectives on human communicative interactions as well.

5. Neural substrates of gesture use

Another interesting recent development is the increasing number of studies that use different brain imaging techniques to investigate (i) whether the use of gestures is lateralized in non-human primates as it is in humans, and (ii) whether this preferential hand use results in an asymmetry of the corresponding brain regions between the two hemispheres (Dadda, Cantalupo & Hopkins 2006; Hopkins, Cantalupo & Taglialatela 2007; Taglialatela, Cantalupo & Hopkins 2006; Taglialatela, Russell, Schaeffer & Hopkins 2008). For example, asymmetries in the chimpanzees' inferior frontal gyrus are only associated with a hand preference for producing POINTING gestures, but not other similar behaviors without a communicative component (Dadda, Cantalupo & Hopkins 2006).

This is particular interesting, since the corresponding human brain region that is homologue to the inferior frontal gyrus in chimpanzees also contains Broca's area. This area is responsible for both the production of language and manual actions, thereby indicating a close link between the two modalities in the human brain.

6. Neglected populations

In their comparison of great apes, siamangs, and Barbary macaques, Call and Tomasello (2007) concluded that monkeys might in general use a smaller amount of manual gestures than apes. However, as mentioned before, still very little is known about gestural communication in monkeys (but see Maestripieri 1996a; 1996b; 1999). Furthermore, many studies on monkeys consider their facial expressions as (facial) gestures (Maestripieri 1999; Paukner, Anderson, Fogassi & Ferrari 2007), while in studies of great apes facial expressions are often considered separately from gestures (Liebal, Pika & Tomasello 2006; Pika, Liebal & Tomasello 2003). There is some recent evidence that Mandrills also use at least a few manual gestures (Laidre 2012) and that baboons use their right hand to produce a manual gesture (Meguerditchian & Vauclair 2006). Thus, there is clearly more research needed to clarify whether monkeys really do use fewer gestures – and here particularly manual gestures – as opposed to the great apes, or whether monkeys have simply not been studied yet to the necessary extent to actually compare gestural complexity of monkeys and great apes.

Furthermore, a frequent critique on gesture research is that most of our knowledge stems from captive populations. As previously mentioned, most of the work on gestural communication has been conducted on great apes in captive settings. There might be different reasons for this gap of knowledge. In the wild, gestures might simply not play such an important role because vocalizations are a more efficient mean to communicate over larger distances in habitats with dense vegetation. An alternative explanation is that to study gestures in wild populations is a very difficult task because of the low visibility in the dense vegetation and since the group is much more dispersed in space. Thus, it remains an open question whether gestures are simply by-products of living in captive settings and whether they actually play a role in interactions of wild populations. Fortunately, there are several papers now published showing that wild chimpanzees indeed use a variety of gestures (Hobaiter & Byrne 2011a, 2012; Pika & Mitani 2008). They use them in referential ways (Pika & Mitani 2006) and also combine them into longer sequences similarly to captive chimpanzees (Hobaiter & Byrne 2011b;

Liebal, Call & Tomasello 2004). Much more research is needed in this regard since studies on gesture combinations are inconsistent in their findings and differ very much in terms of the kinds of definitions and analyses used (Genty & Byrne 2010; Tanner 2004; Tempelmann & Liebal this volume).

7. Future directions and developments

We hope that this chapter provided an overview of current foci of gesture research particularly in non-human primates. We want to close with some suggestions for future research on primate communication.

7.1 What is a gesture?

The focus of our book published in 2007 particularly concerned the question whether gestural communication of non-human primates can be at all compared to gestures used by humans – be it either gestures used by pre-verbal children or gestures accompanying speech (Liebal, Müller & Pika 2007). Although a variety of research projects is now using a comparative approach creating a flourishing field of research into human communication to explore its origins and the forces driving its evolution, there are very few studies that actually use a systematic comparative approach to study communication in both humans and their closest relatives, largely because of the lack of clear criteria of what a gesture actually is. Thus, although there is an increasing body of knowledge, the question of whether gestures of non-human primates share the same properties as human gestures in terms of form, function, repertoire, and use remains unsolved. Thus, what is still needed is a definition with a set of criteria, which is applicable across species (see Scott & Pika this volume). Thus, what a gesture actually is seems still not clearer than when we asked this question some years ago.

7.2 How are gestures acquired?

In addition, more systematic longitudinal studies are needed to investigate how great apes and children – apart from their POINTING gestures – acquire their gestures. A multimodal approach is important to actually understand the complexity of primates' interactions in their social groups and to understand the origin of gestures both in terms of phylogenetic and ontogenetic trajectories. Here, the development of new methods is essential to integrate data collection and analysis for gestural, facial, and vocal means of communication (Slocombe, Waller & Liebal 2011).

7.3 How does the context shape communication?

As the current chapter by Leavens (this volume) indicates, there is still much debate as to whether apes POINT spontaneously without any training, whether they understand the meaning of this distinct gesture, and whether they POINT not only to get something from another individual, but to communicate about external things or events in their environment or even inform others (see also Leavens, Hopkins & Bard 2005; Tomasello 2006; Racine this volume). We think that in general the importance of the raising history and particularly being raised by humans is not considered to the necessary extent, since apes interact with their conspecifics in very different contexts and settings as they do with humans (see Pika & Zuberbühler 2007; Tempelmann & Liebal this volume). Thus, studies should address a systematic comparison of non-human primates' gesture use in their interactions with conspecifics as opposed to interactions with a human experimenter. That the different partners and settings might have an influence is indicated by Hopkins and colleagues (this volume), who found that great apes' preferential hand use while gesturing is more pronounced in interactions with conspecifics.

Thus, the conclusions that can be drawn from a comparative approach to primate gestural communication very much depend on the design of the study, such as the context of interactions (with conspecifics or humans, wild or captive populations of non-human primates, cultural backgrounds of human populations), the methods used for data collection (observations or experiments, continuous or behavior sampling), the raising history of the individuals, as well as which parameters are actually compared between species (e.g. gestures of pre-verbal children to gestures of non-linguistic apes, structural properties or function of gestures).

References

Bard, Kim A. 1992. "Intentional behavior and intentional communication in young free-ranging orangutans." *Child Development* 63 (5): 1186–1197.

Bates, Elizabeth (ed.). 1976. *Language and Context: The Acquisition of Pragmatics*. New York, NY, US: Academic Press.

Bates, Elizabeth (ed.). 1979. *The Emergence of Symbols: Cognition and Communication in Infancy*. New York, NY, US: Academic Press.

Blake, Joana, Vitale, Grace, Osborne, Patricia and Olshansky, Esther. 2005. "A cross-cultural comparison of communicative gestures in human infants during the transition to language." *Gesture* 1 (2): 201–217.

Butterworth, George E. 2003. "Pointing is the royal road to language for babies." In *Pointing: Where Language, Culture, and Cognition Meet*, Sotaro Kita (ed.), 9–34. Mahwah, New Jersey: Lawrence Erlbaum Associates.

Call, Josep and Tomasello, Michael (eds). 2007. *The Gestural Communication of Apes and Monkeys*. New York: Lawrence Erlbaum Associates.

Camaioni, Luigia. 1993. "The development of intentional communication: A re-analysis." In *New Perspectives in Early Communicative Development*, Jacqueline Nadel and Luigia Camaioni (eds 82–96). London: Routledge.

Chevalier-Skolnikoff, Suzanne. 1973. "Visual and tactile communication in *Macaca arctoides* and its ontogenetic development." *American Journal of Physical Anthropology* 38 (2): 515–518.

Dadda, Marco, Cantalupo, Claudio and Hopkins, William D. 2006. "Further evidence of an association between handedness and neuroanatomical asymmetries in the primary motor cortex of chimpanzees (*Pan troglodytes*)." *Neuropsychologia* 44 (12): 2582–2586.

Franco, Fabia and Butterworth, George. 1996. "Pointing and social awareness: Declaring and requesting in the second year." *Journal of Child Language* 23 (2): 307–336.

Genty, Emilie, Breuer, Thomas, Hobaiter, Catherine and Byrne, Richard W. 2009. "Gestural communication of the gorilla (*Gorilla gorilla*): Repertoire, intentionality and possible origins." *Animal Cognition* 12 (3): 527–546.

Genty, Emelie and Byrne, Richard W. 2010. "Why do gorillas make sequences of gestures?" *Animal Cognition* 13 (2): 287–301.

Goldin-Meadow, Susan. 2002. "Constructing communication by hand." *Cognitive Development* 17 (3–4): 1385–1405.

Goldin-Meadow, Susan (ed.). 2003. *The Resilience of Language: What gesture creation in deaf children can tell us about how all children learn language*. New York: Pychology Press.

Gómez, Juan-Carlos. 1990. "The Emergence of intentional communication as a problem-solving strategy in the gorilla." In *"Language" and Intelligence in Monkeys and Apes: Comparative developmental perspectives*, Sue Taylor Parker and Kathleen Rita Gibson (eds), 333–355. Cambridge: Cambridge University Press.

Grigor'eva, O. M. and Deriagina, M. A. 1987. "Gestural forms of communication in primates: I. Development of gestural communication in ontogenesis and phylogenesis." *Biologicheskie Nauki* 1: 45–50.

Halina, Marta and Rossano, Federico. 2010. *The ritualization of gestures in mother-infant bonobos*. Paper presented at the 4th Conference of the International Society for Gesture Studies (ISGS), Frankfurt (Oder), Germany.

Hobaiter, Catherine and Byrne, Richard W. 2011a. "The gestural repertoire of the wild chimpanzee." *Animal Cognition* 14: 745–767.

Hobaiter, Catherine and Byrne, Richard W. 2011b. "Serial gesturing by wild chimpanzees: Its nature and function for communication." *Animal Cognition* 14(6): 1–12.

Hobaiter, Catherine and Byrne, Richard W. 2012. "Gesture use in consortship. Wild chimpanzees' use of gesture for an 'evolutionarily urgent' purpose." In *Developments in Primate Gesture Research*, Simone Pika and Katja Liebal (eds), 127–144. Amsterdam/Philadelphia: John Benjamins Publishing Company.

Hopkins, William D., Cantalupo, Claudio and Taglialatela, Jared P. 2007. "Handedness is associated with asymmetries in gyrification of the cerebral cortex of chimpanzees." *Cerebral Cortex* 17 (8): 1750–1756.

Johnson, Christine M., Zastrow, Shae Lynn and Halina, Marta. 2011. *The socio-ecology of gesture development in bonobos*. Paper presented at the SRCD Biennial Meeting, Montreal, Canada.

Kendon, Adam. 2000. "Language and Gesture: Unity or Duality?" In *Language and Gesture*, David McNeill (ed.), 47–63. Cambridge: Cambridge University Press.

King, Barbara J. and Shanker, Stuart G. 2003. "How can we know the dancer from the dance: The dynamic nature of African great ape social communication." *Anthropological Theory* 3 (1): 5–23.

Laidre, Mark E. 2005. "Honest signaling of intentions: The importance of performance risk, receiver response, and prior knowledge of opponents in the threat displays of mandrills." *Behaviour* 142 (4): 455–476.

Laidre, Mark E. 2008. "Do captive mandrills invent new gestures?" *Animal Cognition* 11 (2): 179–187.

Laidre, Mark E. 2011. "Meaningful gesture in monkeys? Investigating whether mandrills create social culture." *PLoS ONE* 6 (2): e14610.

Laidre, Mark E. 2012. "Mandrill visual gestures: A round-the-world study of the largest of all monkeys." In *Developments in Primate Gesture Research*. Simone Pika and Katja Liebal (eds), 6: 111–126. Amsterdam/Philadelphia: John Benjamins Publishing Company.

Leavens, David A., Hopkins, William D. and Bard, Kim A. 2005. "Understanding the point of chimpanzee pointing." *Current directions in Psychological Science* 14 (4): 185–189.

Liebal, Katja, Call, Josep and Tomasello, Michael. 2004. "Use of gesture sequences in chimpanzees." *American Journal of Primatology* 64 (4): 377–396.

Liebal, Katja, Müller, Cornelia and Pika, Simone (eds). 2007. *Gestural Communication in Nonhuman and Human Primates*. Amsterdam/Philadelphia: John Benjamins Publishing Company.

Liebal, Katja, Pika, Simone and Tomasello, Michael. 2006. "Gestural communication of orangutans (*Pongo pygmaeus*)." *Gesture* 6 (1): 1–38.

Liebal, Katja, Bressem, Jana and Müller, Cornelia. 2010. *Recurrent forms and contexts: Families of gestures in non-human primates*. Paper presented at the 4th Conference of the International Society of Gesture Studies (ISGS).

Liebal, Kristin, Behne, Tanya, Carpenter, Malinda and Tomasello, Michael. 2009. "Infants use shared experience to interpret pointing gestures." *Developmental Science* 12 (2): 264–271.

Liszkowski, Ulf, Carpenter, Malinda, Striano, Tricia and Tomasello, Michael. 2006. "12- and 18-month-olds point to provide information for others." *Journal of Cognition and Development* 7 (2): 173–187.

Liszkowski, Ulf, Brown, Penelope, Callaghan, Tara C., Takada, Akira and De Vos, Connie. in press. "A prelinguistic gestural universal of human communication." *Cognitive Science*.

Maestripieri, Dario. 1996a. "Gestural communication and its cognitive implications in pigtail macaques (*Macaca nemestrina*)." *Behaviour* 133 (13–14): 997–1022.

Maestripieri, Dario. 1996b. "Social communication among captive stump-tailed macaques (*Macaca arctoides*)." *International Journal of Primatology* 17 (5): 785–802.

Maestripieri, Dario. 1999. "Primate social organization, gestural repertoire size, and communication dynamics: A comparative study of macaques." In *The Evolution of Language: Assessing the evidence from nonhuman primates*, Barbara J. King (ed.). Santa Fe: School of American Research.

Marentette, Paula and Nicoladis, Elena. 2012. "Does ontogenetic ritualization explain early communicative gestures in human infants?" In *Developments in Primate Gesture Research*, Simone Pika and Katja Liebal (eds), 33–53. Amsterdam/Philadelphia: John Benjamins Publishing Company.

McNeill, David (ed.). 1992. *Hand and Mind: What gestures reveal about thought*. Chicago: The University of Chicago Press.

Meguerditchian, Adrian and Vauclair, Jacques. 2006. "Baboons communicate with their right hand." *Behavioural Brain Research* 171: 170–174.
Mehrabian, Albert (ed.). 1981. *Silent messages: Implicit communication of emotions and attitudes.* Oxford, England: Wadsworth Pub. Co.
Paukner, Annika, Anderson, James R., Fogassi, Leonardo and Ferrari, Pier F. 2006. "Do facial gestures, visibility or speed of movement influence gaze following responses in pigtail macaques?" *Primates* 48 (3): 241–244.
Perlman, Marcus, Tanner Joanne E. and King, Barbara J. 2012. "A mother gorilla's variable use of touch to guide her infant: Insights into iconicity and the relationship between gesture and action." In *Developments in Primate Gesture Research*, Simone Pika and Katja Liebal (eds), 55–71. Amsterdam/Philadelphia: John Benjamins Publishing Company.
Pika, Simone and Liebal, Katja. 2012. "Introduction: Developments in primate gesture research." In *Developments in Primate Gesture Research*, Simone Pika and Katja Liebal (eds), 1–11. Amsterdam/Philadelphia: John Benjamins Publishing Company.
Pika, Simone, Liebal, Katja and Tomasello, Michael. 2003. "Gestural communication in young gorillas (Gorilla gorilla): Gestural repertoire, learning, and use." *American Journal of Primatology* 60: 95–111.
Pika, Simone and Mitani, John C. 2006. "Referential gesturing in wild chimpanzees (*Pan troglodytes*)." *Current Biology* 16 (6): 191–192.
Pika, Simone and Mitani, John C. 2009. "The *directed scratch*: Evidence for a referential gesture in chimpanzees?" In *The Prehistory of Language*, Rudie Botha and Chris Knight (eds), 166–180. Oxford: Oxford University Press.
Pika, Simone and Zuberbühler, Klaus. 2007. "Social games between bonobos and humans: Evidence for shared intentionality?" *American Journal of Primatology* 70 (3): 207–210.
Plooij, Frans X. 1979. "How wild chimpanzee babies trigger the onset of mother-infant play." In *Before Speech: The beginning of interpersonal communication*, Margaret Bullowa (ed.), 223–243. Cambridge: Cambridge University Press.
Russon, Anne, Andrews, Kristin. 2011. "Orangutan pantomime: Elaborating the message." *Biology Letters* 7 (4): 627–630.
Salomo, Dorothee and Liszkowski, Ulf. 2010. Caretaker-infant social interactions and gestures across cultures. Paper presented at the International Conference on Infant Studies (ICIS), Baltimore, USA.
Sandler, Wendy. 2009. "Symbiotic symbolization by hand and mouth in sign language." *Semiotica* 2009 174: 241–275.
Schneider, Christel, Call, Josep and Liebal, Katja. in press. "What role do mothers play in the gestural acquisition of *Pan pansicus* and *Pan troglodytes*?" *International Journal of Primatology.*
Schneider, Christel, Call, Josep and Liebal, Katja. 2011. "The onset and early use of gestural communication in nonhuman great apes." *American Journal of Primatology* 73: 1–12.
Slocombe, Katie E., Waller, Bridget M. and Liebal, Katja. 2011. "The language void: The need for multimodality in primate communication research." *Animal Behaviour* 81 (5): 919–924.
Taglialatela, Jared P., Cantalupo, Claudio and Hopkins, William D. 2006. "Gesture handedness predicts asymmetry in the chimpanzee inferior frontal gyrus." *NeuroReport* 17 (9): 923–927.
Taglialatela, Jared P., Russell, Jamie L., Schaeffer, Jennifer A. and Hopkins, William D. 2008. "Communicative signaling activates 'Broca's' homolog in chimpanzees." *Current Biology* 18 (5): 343–348.

Tanner, Joanne E. 2004. "Gestural phrases and gestural exchanges by a pair of zoo-living lowland gorillas." *Gesture* 4 (1): 1–24.

Tempelmann, Sebastian and Liebal, Katja. 2012. "Spontaneous use of gesture sequences in orangutans: A case for strategy?" In *Developments in Primate Gesture Research*, Simone Pika and Katja Liebal (eds), 6: 73–92. Amsterdam/Philadelphia: John Benjamins Publishing Company.

Tomasello, Michael. 2006. "Why don't apes point?" In *Roots of Human Sociality: Culture, cognition, and interaction*, Nicholas J. Enfield and Stephen C. Levinson (eds), 506–524. Oxford & New York: Berg Publishers.

Tomasello, Michael, Call, Josep, Nagell, Katherine, Olguin, Raquel and Carpenter, Malinda. 1994. "The learning and use of gestural signals by young chimpanzees: A trans-generational study." *Primates* 35 (2): 137–154.

Tomasello, Michael, Call, Josep, Warren, Jennifer, Frost, G. Thomas, Carpenter, Malinda and Nagell, Katherine. 1997. "The ontogeny of chimpanzee gestural signals: A comparison across groups and generations." *Evolution of Communication* 1 (2): 223–253.

Tomasello, Michael, George, Barbara L., Kruger, Ann Cale, Farrar, Michael Jeffrey and Evans, Andrea. 1985. "The development of gestural communication in young chimpanzees." *Journal of Human Evolution* 14 (2): 175–186.

Tomasello, Michael, Gust, Deborah and Frost, G. Thomas. 1989. "A longitudinal investigation of gestural communication in young chimpanzees." *Primates* 30 (1): 35–50.

Wilkonson, Ray, Leudar, Ivan, and Pika, Simone. 2012. "Requesting behaviours within episodes of active sharing: A new look on chimpanzee signalling." In *Developments in Primate Gesture Research*. Simone Pika and Katja Liebal (eds), 197–218. Amsterdam/Philadelphia: John Benjamins Publishing Company.

van Hooff, Jan A. R. A. M. 2012. "Facial-vocal displays, gestures and language." In *Developments in Primate Gesture Research*. Simone Pika and Katja Liebal (eds), 13–31. Amsterdam/Philadelphia: John Benjamins Publishing Company.

Index